Fisheries Management in Japan

FISH & FISHERIES SERIES

Volume 34

Series Editor: David L.G. Noakes, Fisheries & Wildlife Department, Oregon State University, Corvallis, USA

For other titles published in this series, go to
www.springer.com/series/5973

Mitsutaku Makino

Fisheries Management in Japan

Its institutional features and case studies

Springer

Mitsutaku Makino
National Research Institute of Fisheries Science
Fisheries Research Agency
Fukuura 2-12-4
236-8648 Yokohama, Kanagawa
Kanazawa-ku
Japan

ISBN 978-94-007-1776-3 e-ISBN 978-94-007-1777-0
DOI 10.1007/978-94-007-1777-0
Springer Dordrecht Heidelberg London New York

Library of Congress Control Number: 2011935459

Printed on acid-free paper

Springer is part of Springer Science+Business Media (www.springer.com)

Foreword

In the spring of 21011 the attention of the world has been focused on Japan and the tragedy unfolding from a major earthquake, a catastrophic tsunami, and the uncertainty of the impacts of those natural disasters on a nuclear generating center. The physical devastation is overwhelming even when viewed from a distance. The human cost is almost beyond our comprehension. All of us with colleagues and friends in Japan rushed to contact them by any means available to try to offer any help or assistance. By coincidence I happened to have three colleagues from Miyagi University in Sendai, Japan, visiting in my laboratory when this happened so the experience was even more direct for us. They could not make any contact with their families or colleagues for several days, and could not make travel connections to return home. We shared their fears, uncertainties, and concerns until they were able to learn that their families and homes had survived. They were not able to return home to Japan for almost 3 weeks. Remarkably they were able to maintain their composure and complete their planned research project before their departure. They did so with dignity, reserve, and total dedication.

Perhaps we should understand this as a statement of the character of Japan and the Japanese people. In a broader historical context we know that Japan has survived numerous upheavals, catastrophes, and disasters. It is the Japanese character to survive, to adapt, to work together, and to persevere. So it is with Japanese fishing and Japanese fisheries. This book could not be more timely or more important to all of us.

We all know something about Japanese fishes and fisheries. Japan is a world leader in harvest, production, and consumption of fishes and fishery products. Japanese scientists represent the state of the art in many areas of fish biology, fisheries, and fisheries technology. Japanese cuisine is regarded as an ultimate combination of culinary and artistic expression. It receives the highest possible compliment in the form of imitation in virtually every country worldwide. Japanese poetry, artwork, and even architecture all have strong influences of fishes and fisheries. We know that Japan has a very long and rich human history, but the connections to fishes and fisheries are not so well known.

This volume is a landmark. It documents the history of Japanese fishing and fisheries management, perhaps for a longer period than for any other country. The detail includes almost every aspect of human and environmental influences that could affect fishes and fishing. The authors provide detailed analyses of selected case studies, including the most current consideration of marine protected areas. Concepts as fundamental as ecosystem-based management and conservation of biodiversity are discussed in the context of the Japanese fisheries system. The authors explain the basis for fisheries management in Japan, and how that has changed over time. Much of this information has not been readily available before – at least not in the English language literature. One immediate impact of the current tragic devastation in Japan is the disruption of fishing and fisheries. It has been noted by some commentators that the resolution of the present situation will require a fundamental reorganization of Japanese fisheries management practices. Japan and Japanese fisheries have survived before and they will survive now. Major social, cultural, economic, and political changes have been imposed upon the country and the people over a long history. The people and the country have changed and adapted in the past – how they will change in the future is not certain. What is certain is that the information and knowledge in this volume will provide the basis for everything that might develop in the future.

Editor, Kluwer Fish and Fisheries Series Dr. David L.G. Noakes
Professor of Fisheries and Wildlife
Oregon State University
Senior Scientist, Oregon Hatchery Research Center
Corvallis, Oregon 97331-3803
USA

Preface and Acknowledgements

Japan is one of the world's largest fish-eating countries with a long history, and has developed its own customs and values in terms of managing fisheries resources. However, the social science works on Japanese fisheries have not been well presented in international publications to date. This volume is the first English book in more than 20 years on the social science aspects of Japanese fisheries. The main objective of this book is to introduce the history and institutional features of capture fisheries management in Japan, with nine case studies from sub-arctic to tropical ecosystems, from sedentary to migratory species, and from small-scale coastal to offshore industrial fisheries.

It is well known that the theory of fisheries management has been led by researchers in, the so-called Western countries, such as the USA, the UK, Canada, Australia, New Zealand, Iceland, Norway, etc. Yet Japan has its own history, its own customs and values with regard to how to manage fisheries resources, as well as fisheries communities. This Japanese approach is considerably different from that of the Western countries. I hope this book will contribute to mutual understanding between the West and the East, and to constructive discussions on the future of fisheries management around the world.

The first part of this book briefly introduces the institutional history of Japanese fisheries management, which spans more than 1,300 years. Of course, the institutional frameworks for fisheries management differ from sector to sector. For example, coastal fisheries management is more community-oriented, and local people have the authority and take priority in the decision-making process. In contrast, offshore fisheries are more industrialized and commercially oriented. The national government plays a principal role in the planning stage of offshore resource management, and the offshore fisheries organizations play a role in the implementation and distribution processes. By examining essential case studies, this book describes the coastal and offshore fisheries managements, and discusses the differences in their nature.

One of the main challenges in world fisheries is to implement the "ecosystem-based approach" to management. There are no one-size-fits-all measures for the ecosystem-based approach. This book investigates the advantages and limitations of the Japanese fisheries management system from the viewpoint of the Convention on Biological Diversity, and discusses the environmental policy measures needed to fill the gaps between fisheries management and ecosystem-based management. As a case study, management measures and administrative costs in the Shiretoko World Natural Heritage area are analyzed. Based on the above analysis, the final part of this book is about the Gland Plan of Japanese fisheries policy for the next 20 years. Here, three policy options and future scenarios on Japanese fisheries are presented.

I would like to acknowledge my heartfelt gratitude for the rigorous academic mentorship of Professors Yoshifusa Kitabatake, Wataru Sakamoto, and Tateki Fujiwara of Kyoto University, Japan. Especially, Professor Masayuki Yamagishi kindly showed me "the art of life" as an academic researcher as well as how to be a good husband and father. I am also grateful to my co-authors on journal papers, books, and reports over the years, for their stimulation and insights. Many of those insights have undoubtedly found their way into this book, although any errors are my own and not theirs; thank you: Hiroyuki Matsuda, Yasunori Sakurai, Minoru Tomiyama, Takumi Mitani, Yuji Machiguchi, Masahito Hirota, and Takaomi Kaneko. I also appreciate the support of Springer, notably Martine van Bezooijen, as well as that of series editor Professor David L.G. Noakes of Oregon State University. I am very grateful to my research assistant, Ms. Shigeko Usami, who has provided invaluable support in the preparation of this book.

Finally, on a personal note, I would like to take this opportunity to thank my parents for their many years of continuous support. Above all, I thank my wife, Mikako Makino, for her patient support of my work.

Fisheries Research Agency
Yokohama, Japan

Mitsutaku Makino

Contents

List of Photo Copyrights

Chapter 1
Introduction

Abstract This chapter gives an outline of Japan and her fisheries. General information on Japan, including geography, ocean currents, population, the structure of the national economy, food culture, etc., is followed by an overview of Japanese fisheries sector that includes the principal legal system, administrative structure, insurance system, and other social institutions linked to the Japanese fisheries sector. Fisheries infrastructure, such as fishing grounds, fishing ports and roads, as well as education and statistic system, is included. A different but equally important fisheries infrastructure is the research and promotion of scientific knowledge to local fishers.

1.1 Japan

1.1.1 Geography

Japan (Nihon or Nippon in Japanese) is a Pacific country in East Asia, composed of 6,852 islands. It has an area of 377,944 km^2, which ranks it as the 60th largest country in the world. Its four major islands are, from north to south, Hokkaido, Honshu ("Main Island"), Shikoku, and Kyushu and Okinawa (Ryukyu) (Fig. 1.1). The Japanese archipelago extends for more than 3,000 km, covering subarctic to tropical ecosystems. Most of the land is mountainous, with numerous dormant and some active volcanoes. Flat areas account for only 4.9% of the total land, most of them located along the coastline. The total length of the coastline is 29,751 km (CIA 2010).

The areas of Japanese territorial waters and the 200-nautical-mile Exclusive Economic Zone (EEZ) are 430,000 and 4,050,000 km^2, respectively. The area of the EEZ is about 11 times the size of the country's land area, and is ranked as the sixth largest in the world, after the USA, Australia, Indonesia, New Zealand, and Canada (Ocean Policy Research Foundation 2004).

M. Makino, *Fisheries Management in Japan: Its institutional features and case studies*,
Fish & Fisheries Series 34, DOI 10.1007/978-94-007-1777-0_1,

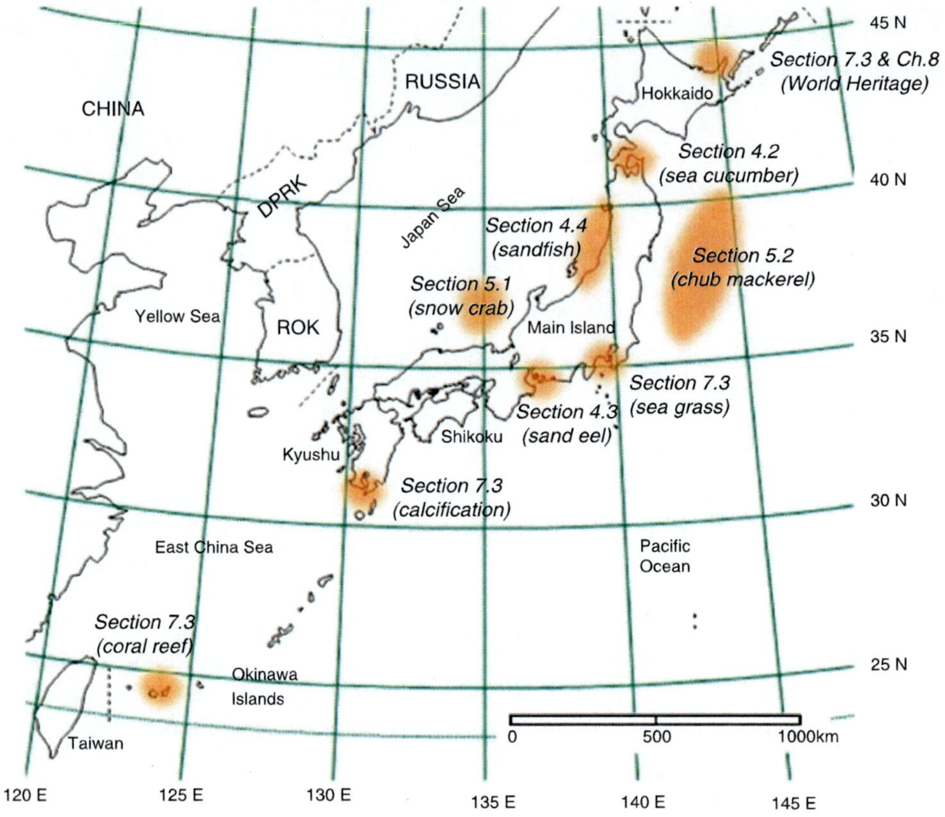

Fig. 1.1 Geography of Japan and cases presented in this book

There are three major ocean currents brushing the Japanese islands. The Kuroshio, or Black Current, is a warm current that originates in the North Equatorial Current and flows northward via the South China Sea. The Kuroshio Current is strongest in spring and summer. Another warm current, the Tsushima Current, is a western branch of the Kuroshio Current that flows northward along the Japan Sea side. The Oyashio Current is a cold current, which originates in the Bering Sea and Okhotsk Sea and flows southward along Japan's Pacific coastline. It is rich in nutrients, and strongest in winter. The Kuroshio and Oyashio collide in the Pacific offshore area of Honshu Island, forming large current rips. This area is one of the most productive fishing grounds in the world.

1.1.2 Population and Economic Structure

The total population of Japan was 127.7 million in 2008. Twenty-two percent of the Japanese are over 65 years old, while only 13.5% are under 14 years old. It is predicted that, by 2055, the total population will have fallen below 90 million and more

Table 1.1 Composition of protein intake for the average Japanese (Source: MAFF 2009)

Food item	Share of total protein intake (%)	Share of animal protein intake (%)
Fisheries products	20.5	37.4
Meats	17.8	32.5
Eggs	7.1	12.9
Milk and dairy products	9.4	17.2
Cereals	24.4	–
Beans	9.3	–
Others	11.5	–

than 40% will be over 65 years old, while fewer than 9% will be under 14 years old (Kaneko et al. 2008). Advancing average age, combined with a falling birthrate, is one of the most serious problems facing Japan.

Another issue relating to the national population is spatial allocation. The average density of the Japanese population in 2005 was 343 people per square kilometer, about the same as that in Belgium (348 in 2001) or India (345 in 2001), smaller than the Netherlands (439 in 2002), and 11 times greater than that of the USA (31 in 2000). However, the Japanese population is highly concentrated in just a few urban areas, such as the Tokyo Metropolitan Area (5,751 people per square kilometer in 2005), Osaka Prefecture (4,655), and Kanagawa Prefecture (3,639). Northern remote areas, such as Hokkaido Prefecture (72), Iwate Prefecture (91), and Yamagata Prefecture (99) are much less populated. In most of these remote areas, the fisheries sector plays a crucial role in the local economies.

The Japanese Gross Domestic Product (GDP) in 2007 was 515.9 trillion yen (Economic and Social Research Institute of Cabinet Office 2009). The three largest sectors in the Japanese economy were the service industry, manufacturing industry, and real estate industry, which produced 22.0%, 21.1%, and 11.9% of total GDP, respectively. Primary industries, i.e., agriculture, forestry, and fisheries, produced 1.2%, 0.1%, and 0.2% of GDP, respectively. In 2007, the total worker population was 66.7 million, with an unemployment rate of 3.8%. Of these, 68.2% were engaged in tertiary industries (service-based industries), while 4.2% were in primary industries, including fisheries. As the above figures show, the fisheries sector is a very small part of the Japanese national economy.

1.1.3 Food Culture

At the end of this section, it is worth mentioning the food consumption habits of the Japanese people. The Japanese are fish-eaters. The per capita supply of seafood was 61.5 kg in 2008. This figure is the second largest in the world after Iceland. The average Japanese eats 80.2 g of protein per day, of which 43.9 g is fisheries products. Table 1.1 shows the composition of the total protein intake for the average Japanese. Fisheries products are the second largest source of total protein intake, and the largest source of animal protein intake.

Figure 1.2 shows an international comparison of the percentage of fisheries products as a source of animal protein in the top 40 fishing countries in the world. The top 40 fisheries countries are defined using average production volume (tons) for 2002–2006, and arranged according to the latitude of their capital city. The Faroe Islands and Taiwan are left out, due to lack of data on animal protein sources. The figure shows that the low-latitude countries in the Asia-Pacific area and African coastal area have a greater reliance on fisheries products than other sources of animal protein, reflecting the importance of seafood in their food security. Also, old fisheries countries in northern Europe, i.e., Iceland and Norway, show higher values than seen in other European countries. Likewise, Korea and Japan show high values among the mid-latitude countries. These four countries are so-called developed countries, which suggests that people living there enjoy and choose fisheries products as a part of their food culture (Makino and Matsuda 2011).

Docosahexaenoic acid (DHA) and eicosapentaenoic acid (EPA), which are polyunsaturated fatty acids found in lipids in fisheries products, prevent lifestyle-related diseases and the formation of blood clots, and help with brain development. In addition, fisheries products are rich in nutrients that may be lacking in other parts of the Japanese diet, such as calcium and iron. Therefore, it is often said that Japanese traditional food culture, which involves consuming a lot of fisheries products, is one reason for the good health and long life expectancy of the Japanese. According to the United Nations World Population Prospects (2007), the world's two longest life expectancies at birth are 82.6 years in Japan and 81.8 years in Iceland.

The five main ways in which fisheries products in Japan are served are as sashimi (raw fish), yaki-zakana (grilled fish), ni-zakana (boiled fish), himono (dried fish), and tempura (deep-fried fish). The Japanese are particularly fond of sashimi. The famous Japanese food, sushi, is a combination of sashimi or vegetables with vinegared rice. The most important factor in the quality of sashimi is its freshness. Yaki-zakana (grilled fish) also requires fresh fish, although it is said that freshness is less important for ni-zakana (boiled fish). Therefore, in the Japanese domestic market, prices of same size and same species differ considerably according to their freshness.

The Japanese eat fish and rice almost every day. The average household spent ¥88,593 on fisheries products in 2008, of which 59% comprised fresh fish (Ministry of Internal Affairs and Communications 2009). An interesting fact is that the average Japanese consumes more fish as they get older. Also, according to comparative studies between areas in Japan, the consumption volumes of fish and rice are positively correlated. People in the northern parts of Japan consume more fish and rice than in other areas (Nagasaki 1994).

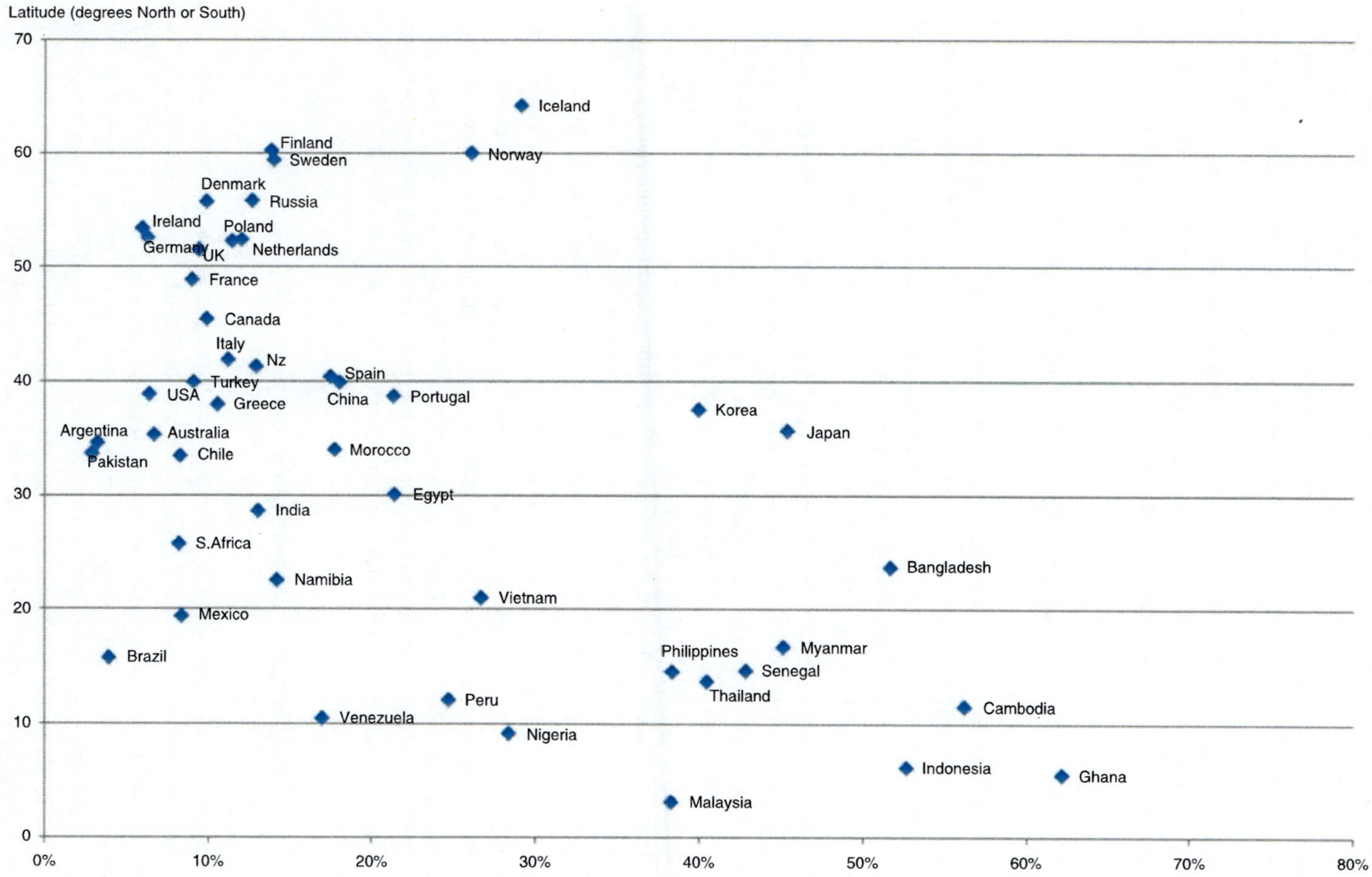

Fig. 1.2 Percentage of fisheries products as a source of animal protein in the top 40 countries with fisheries, arranged by latitude of their capital city (Source: FAO 2004–2008)

1.2 Overview of Japanese Fisheries

Japanese fisheries target a wide range of species using a wide variety of methods and equipment. The administrative system also has a complicated way of categorizing types of fishing practices. To start with, there are three ways of classifying Japanese fisheries and their profiles. These are classifications according to fishing equipment, to fishing grounds, and based on administrative definitions. More details on the current state of the Japanese fisheries sector will be presented in Chap. 3.

1.2.1 Classification According to Fishing Equipment

Due to the long history of fisheries operations conducted all along the Japanese islands, a wide range of equipment and methods have been developed in Japan. There are also many similar types of equipment in various parts of Japan with a variety of local names. Theoretically, fishing equipment used in Japan is categorized into the following three categories (Asada et al. 1973; FAO 1993; Kaneda 1995):

1. Net fishing: bottom trawl fishing, (such as Danish seines, dredge nets, otter trawls, pair trawls, etc.), boat seine fishery or mid-water trawl fishing, beach seine fishing, purse seine fishing, blanket net fishery (such as four-arm scoop-nets, stick-held dip nets, multi-boat lift nets, etc.), gillnet fishing, set-net fishing, etc.
2. Angling fishery: handline fishing, pole and line fishing, mechanical angling fishing, trolling line fishing, vertical longline fishing, and longline fishery.
3. Other fisheries: baitless angling fishing, brush weir fishing, trap fishing, shelter fishing, diving apparatus fishing, spear and dart fishing, hook fishing, etc.

For example, squid are captured by trap fishing (cages), angling (jigging), driftnet fishing, purse seine fishing, set-net fishing, trawl fishing, etc. Therefore, we don't use the species name to describe a fisheries sector (e.g., squid fishery) because they are caught with too many types of fishing equipment. When a specific fisheries sector is described according to its equipment, we often use a combination of species name and equipment name, such as "squid jigging fishing" or "tuna longline fishing."

1.2.2 Classification by Fishing Grounds

Another type of categorization is the location of fishing grounds, or the distances of fishing grounds from the coastline. There are four of these types of fisheries: coastal fisheries, offshore fisheries, distant-waters (high-seas) fisheries, and inland water fisheries. This classification often coincides with the size of the vessels used in the sector, making it convenient for informal use. However, no official definition exists for these three fisheries in Japan, possibly because how these fisheries are generally viewed changes over time. They are currently viewed as follows:

1. Coastal fisheries operate for 1 or 2 days at sea per single operation and are the biggest in terms of production value and employment. They take very many

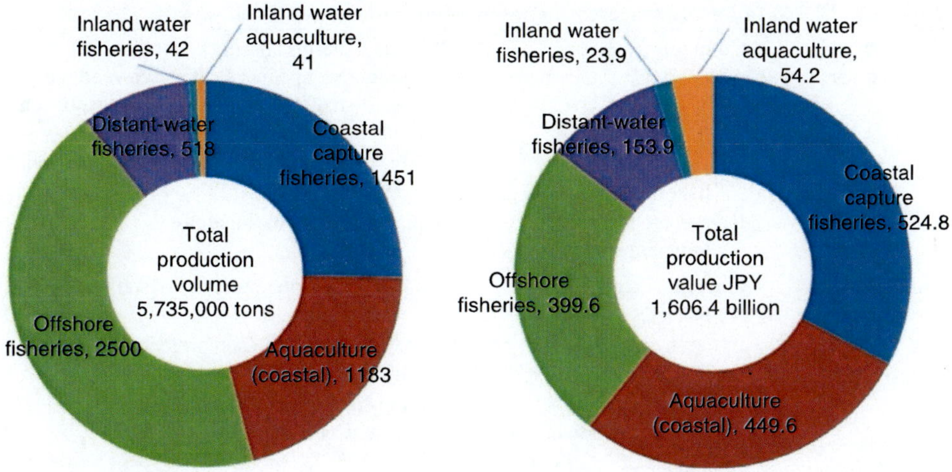

Fig. 1.3 Production volume and value from each fisheries sector in 2006 (Source: Fisheries Agency 2009)

forms, and marine aquaculture is usually included in this category. Because the fishing grounds are close to the shore, most of these fisheries are managed by local organizations of fishers, such as Fisheries Cooperative Associations (FCAs). Fishers from other areas are not permitted to operate there, or need to obtain official approval from local FCAs if they want to. The chief target species are sedentary or local species, although large numbers of migratory finfish species are captured as well. Coastal fishers use small-scale vessels. The fisheries annual statistics published by MAFF (Ministry of Agriculture, Forestry and Fisheries) define coastal fisheries as fisheries using vessels of less than 10 gross tons, but in practice they are up to about 15 gross tons. Most of these fisheries businesses are run by family members. Rights-based fisheries (explained later) are all categorized in this type of fisheries.

2. Offshore fisheries are industrialized fisheries, and the most important sector in terms of production volume. The normal size of their vessels is up to about 100 gross tons, although there are many exceptions. Offshore fisheries vessels move across prefectural borders and operate widely within the EEZ. Because their volume-wise catching efficiency is much higher than coastal fisheries, there have been a lot of conflicts with coastal fisheries. For example, before the coastal fisheries harvest matured individuals that are migrating toward coastal areas to spawn, offshore fisheries can capture large numbers of immature individuals in fishing grounds close to the coastal area. Therefore, one of the most important tasks for the government (mainly the national government) is to coordinate these two fisheries.
3. Distant-waters fisheries are highly industrialized setups operating on the high seas or the EEZs of other countries. As will be described in Chap. 3, it was the most productive sector in Japan before the establishment of the 200 nautical miles regime worldwide. The production of these fisheries has been gradually shrinking.
4. Inland water fisheries are very small in production value and volume in Japan. Figure 1.3 shows the production volume and value from each sector in 2006.

Table 1.2 Classification of Japanese fisheries by administrative definition

Rights-based fisheries	Common fisheries	Various fisheries sharing coastal areas. Sedentary species such as seaweed, kelp, shellfish, etc., small-scale set-nets, beach seines, and inland waters
	Large-scale set-net fisheries	Using set-nets over 27 m in depth in coastal areas
	Demarcated fishery (aquaculture)	Aquaculture in coastal areas
License-based fisheries	Governor-licensed fisheries	Medium-scale operators in coastal and offshore areas
		Numerous types in each prefecture depending on local histories and customs
	Minister-licensed (designated) fisheries	Large-scale operators offshore and in high-seas areas
Other fisheries	Free fisheries, notified fisheries, etc.	

Table 1.3 Minister-licensed (designated) fisheries in 2008

Name of fishery license	Number of vessels in operation	Catch in 2008 (1,000 ton)
Offshore bottom trawl fishery	387	415
East China Sea bottom trawl fishery	13	7
Distant-waters bottom trawl fishery	44	63
Large- and medium-scale purse seine fishery	206	860
Large-scale whaling	0	0
Small-scale whaling	9	103 head (Baird's beaked and Pilot whales)
Mother-ship-type whaling	0	0
Distant-water tuna fishery	567	206
Near-water tuna fishery	434	105
Middle-scale salmon driftnet fishery	59	5
North Pacific saury fishery	197	269
Japan Sea red queen crab fishery	15	10
Squid jigging fishery	177	73

(Source: Fisheries Yearbook Editorial Committee 2009)

1.2.3 Classification by Administrative Definition

As will be detailed in Sect. 1.4, Japanese fisheries can be categorized into three administrative types: rights-based fisheries, license-based fisheries, and "other" fisheries (Table 1.2). The most industrialized fisheries are the Minister-licensed (Designated) Fisheries. Based on Article 52 of the Fisheries Law, there were 13 Minister-licensed Fisheries in 2009 (Table 1.3). The current states of major fisheries sectors defined by the above classifications are presented in Sect. 3.2.

1.3 Social Institutions Linked to Fisheries

1.3.1 Legal Framework

The principal law for fisheries productions in Japan is the Fisheries Law of 1949 (as amended). This law governs administrative categories (rights and licenses in Table 1.2) and other regulations on fisheries. Also, the Fisheries Law defines various levels of fisheries-coordinating organizations. Details of the fisheries management system under this law will be presented in Sect. 2.3.

Another law that is closely related to the Fisheries Law is the Fisheries Cooperative Associations Law of 1948. This law governs the fishers's and fisheries processors' organizations. The Fisheries Cooperative Associations (FCAs), which play the core roles in fisheries management, as described in case studies later on in this book, are organized based on this law. There are more than 1,000 FCAs in Japan.

The Fisheries Resource Protection Law of 1951 defines various regulations for the protection of fisheries resources. Most of the official regulations for fisheries operations issued from central and local governments are based on this law.

The Fishing Vessel Law of 1950 governs the construction, registration, inspections, etc., of vessels used in fisheries operations. The Law for Fishing Ports and Fishing Ground Construction of 1950 defines the planning and maintenance of fishing ports and fishing grounds by the government. It also administrates the environmental restoration programs or ecosystem conservation activities, as well as the construction of infrastructure in fishing communities such as ports or roads.

The above laws came into force during the period of the Fisheries Reform after the WWII, under the heavy influence of the Allied Powers (i.e., the United States of America). More details on the background and legislative histories of these laws will be presented in Sect. 2.3.

Another important law governing resource management is the Law Regarding the Preservation and Management of Living Marine Resources of 1996. This law was enacted after Japanese ratification of the United Nations Convention on the Law of the Sea (UNCLOS). Based on this law, the Total Allowable Catches (TACs) system was introduced as the very first output-control measure for Japanese fisheries.

Finally, there are two Acts that provide the basic policy framework for fisheries. The first is the Basic Act on Fisheries Policy of 2001. This act sets out the two principles of Japanese fisheries policy, which are (1) securing a stable supply of fishery products and (2) appropriate development of fisheries sectors. The other is the Basic Act on Ocean Policy of 2007. This act provides a framework for coordinating various sectors relating to the oceans, such as fisheries, ship transportation, marine defense, the marine environment, marine science, etc. Details of these relatively new systems are presented in Sect. 2.4.

1.3.2 Administrative Structure

The national administrative body for the fisheries sector is the Fisheries Agency (FA) of the Ministry of Agriculture, Forestry and Fisheries (MAFF). There are four departments and 14 divisions within the FA: the Fisheries Policy Planning Department (Administrative Division, Policy Planning Division, Fisheries Management Improvement Division, Fisheries Processing and Marketing Division), the Resource Management Department (Resource Management Division, Fisheries Coordination Division, Far Seas (distant waters) Fisheries Division, and International Affairs Division), the Resource Enhancement Promotion Department (Research and Technological Guidance Division, Resources and Environment Research Division, and Fish Ranching and Aquaculture Division), and the Fisheries Infrastructure Department (Planning Division, Construction Division, and Fishing Communities Promotion and Disaster Prevention Division). The Fisheries Agency also has six local offices, called the Fisheries Coordination Offices. Their locations and the areas they are in charge of are the Hokkaido office (for areas around Hokkaido Island), the Sendai office (for the Pacific Ocean), the Niigata office (for northern parts of the Japan Sea), the Sakai-minato office (for the western parts of the Japan Sea), the Setouchi office (for the Seto Inland Sea), and the Kyushu office (around Kyushu Island). About 900 personnel are working at the Fisheries Agency and the local offices on preserving and managing marine biological resources and fishery production activities, and coordinating at the national and international level. The FA's annual budget in 2008 was ¥242 billion.

In Japan, the marine waters are divided into a number of sea areas as administrative units, and a good many administrative tasks, especially relating to coastal fisheries, have been delegated to the coastal prefectural governments. Therefore, almost all prefectures have a department of fisheries administration. For example, fishing rights, which apply only to coastal areas, and fishery licenses for coastal areas are granted or issued by the governor of each prefecture. Administrative tasks relating to local fisheries cooperation are mostly taken care of at the prefectural level.

1.3.3 Insurance and Finance Systems

The fishing industry faces a variety of specific uncertainties, such as resource fluctuations, volatile fish and fuel prices, natural disasters like typhoons or tsunamis, environmental pollution, red tides, the outbreak of fish diseases, etc. Likewise, fishery business management entails many risks, such as accidents in fishing operations and the ill health of fishers. Mutual insurance schemes are effective in hedging against these uncertainties and promoting the development of the fishing industry. Financial institutions specializing in fishery loans and corresponding debt guarantee systems have also been developed to deal with the above risks.

Table 1.4 Mutual relief insurance schemes for fisheries in Japan

Scheme	Function
Mutual Relief for Catch Security	Compensation for losses resulting from reduced catch value due to poor catch, etc. (harvest insurance)
Mutual Relief for Aquaculture	Compensation for losses resulting from the death or destruction of cultured aquatic animals and plants (damage insurance)
Mutual Relief for Specified Aquaculture	Compensation for losses in specified aquaculture operations resulting from disease, pests, etc. (harvest insurance)
Mutual Relief for Fishery Facilities	Compensation for losses resulting from damage to fishing gear or aquaculture facilities (damage insurance)

1.3.3.1 Mutual Relief Insurance Schemes

Mutual relief insurance for fisheries is designed to contribute to the stability of fish production and business management by compensating for losses resulting from unexpected phenomena or accidents such as natural stock fluctuations, abnormal sea conditions, typhoons, and tsunamis. As shown in Table 1.4, four types of nonprofit mutual relief insurance schemes have been established in Japan under the Fisheries Cooperative Association Law of 1948. A proportion of the insurance premiums are subsidized by the national government. The system of mutual relief insurance comprises local Fisheries Cooperative Associations (FCAs), Prefectural Mutual Relief Associations that are formed at the prefectural level, and the National Federation of Fisheries Mutual Relief Associations, which acts as the governing body.

1.3.3.2 Fisheries Finance

Lending from private financial institutions is hard to come by for fishers, largely because the fishing industry, by nature, is susceptible to natural conditions and fluctuation in income, requires a substantial amount of initial investment and a long payout time, and may only pledge security assets of atypical character. Therefore, in addition to market-based private financing, policy-based financing from the central or local government has been developed. In Japan, three types of financial institutions play a major part in fisheries finance.

Major government-affiliated financial institutions related to fisheries include the Development Bank of Japan and the Okinawa Development Finance Corporation, a public institution designed to promote industrial development in Okinawa Prefecture. Many of the loans are provided for policy purposes, including compensation required for fleet reduction, the construction of fishing vessels or installation of aquaculture facilities intended to develop remote fishing villages, the development of infrastructure, including fishing ports and fishing grounds, and the introduction of specific fishing vessels or facilities.

As a nonprofit finance system, the Fisheries Cooperative Credit System comprises local FCAs, their prefectural federations (Federations of Fisheries Credit Cooperatives), and the Central Cooperative Bank for Agriculture and Forestry. Backed by the FCA savings scheme, the system serves as the mainstay of Japanese fisheries finance, accounting for 66.7% of outstanding fisheries-related loans with a deposit balance of ¥909 billion as at the end of fiscal 2007 (Fisheries Yearbook Editorial Committee 2009).

To facilitate lending from financial institutions, the Fishery Credit Guarantee Associations provide guarantees for loans. Organized at the prefectural level, the members of these associations include prefectural and municipal governments, and medium- and small-scale fishers. At the national level, the Agriculture, Forestry and Fisheries Credit Foundation provides reinsurance for prefectural corporations.

A member may obtain a guarantee by paying an annual charge equivalent to 0.25–1.5% of the loan amount to be guaranteed. A Fishery Credit Guarantee Association may secure reinsurance by paying to the national foundation an annual premium equivalent to 0.22–1.20% of the loan amount to be guaranteed.

1.4 The Fisheries Infrastructure

The fisheries infrastructure is important for establishing the stable provision of fisheries products to Japanese citizens and ensuring the development of the fishing industry. This infrastructure includes fishing grounds with good habitats for plants and animals, fishing ports and access roads thereto, meeting facilities to serve as fora for fishery operators' decision making, wastewater treatment facilities, etc. Typically, this type of infrastructure is available to multiple users at the same time. On the other hand, it tends to be difficult for such facilities, once established, to collect fees in relation to their extent of use or to deny access to any specific user or group. They are characterized as public goods. It is therefore incumbent on the government to develop this infrastructure as a public project. There are two types of fisheries infrastructure in Japan: hard infrastructure and soft infrastructure.

1.4.1 Hard Infrastructure

There are three main constituents of the hard aspects of infrastructure: fishing grounds, fishing port facilities, and fishing villages. Japan is currently engaged in hard infrastructure development projects based on the Act on the Development of Fishing Ports and Grounds of 2001. The total budget for fiscal year (FY) 2009 was ¥162 billion. Before pursuing hard infrastructure projects, however, it is of primary importance to fully weigh the balance of expected costs and benefits, to address maintenance costs after establishment, and to create flexible plans that are reviewed regularly after identifying project outcomes.

The most fundamental infrastructure for the fishing industry is the fishing grounds where marine plants and animals can live and breed. The Fishing Grounds Environment Improvement Program is a set of government initiatives to create this infrastructure. This is the most central and essential program among all the infrastructure-building programs to promote fisheries. The priorities under this program are the protection and restoration of sea grass beds, tidelands, and coral reefs, which are being lost because of industrial development and environmental pollution. The creation of spawning and breeding reefs and the removal of silt to improve water quality are also prioritized in this program. Reforestation and forest conservation projects are also being implemented in concert with the Forestry Agency of MAFF and the forestry departments of local governments in the upper reaches of rivers that flow into fishing grounds.

Another important infrastructure is fishing ports. Fishing ports serve as both the location where harvested fish are landed and the starting point for the distribution of landed products to consumers. These ports also need to protect fishing vessels and equipment from typhoons, tsunamis, and other natural disasters. For such purposes, some public funds are used to develop breakwaters, docks, slipways for ship maintenance, fueling stations, fishing gear warehouses, landing points for fishery products, refrigeration facilities, water services, and icemakers as well as waiting rooms, meeting rooms, and office facilities for fishers. Priority is also given to initiatives to beef up quality control and hygiene management at fishing ports, including through the introduction of systematic food safety programs such as HACCP (Hazard Analysis and Critical Control Points) and traceability systems.

The fishing villages surrounding fishing ports serve as living spaces for fishers and related business operators and their families. At the same time, the garbage and wastewater produced by fishing villages risk causing environmental degradation in nearby fishing grounds. This makes it important to improve the residential environment in these settlements, not only to contribute to the development of fishing villages and the fishing industry as a whole, but also because it raises the living standards of fishing village inhabitants. Through a number of infrastructure-development programs, the Fisheries Agency is involved in the construction of roadways, electricity, water services, wastewater treatment facilities, garbage recycling facilities, etc.

1.4.2 Soft Infrastructure

The soft infrastructure means knowledge gained through fisheries-related research and education and the creation of a fisheries statistics system. Fisheries-related research activities are dealt with separately in the next subsection: this subsection gives an outline of fisheries-related education and fisheries statistics in Japan.

1.4.2.1 Human Resource Development Through Fisheries Education

In general, fisheries education plays a vital role in cultivating leaders and instructors who spur on the fishing industry. While the format of such education varies according to the circumstances and directions of the fishing industry in each country and region, the need for workers with special skills and knowledge is on the rise everywhere as the industry becomes more sophisticated. Fishery high schools and universities have, over the generations since the 1880s, helped to meet the labor needs of the Japanese fishing industry and support its development.

In Japan, the first fishery high schools were established in 1894 for the purpose of training mid-level technicians for the industry. As of 2008, there were 46 fishery high schools or high schools with fisheries programs across the country, with 10,377 students (about 23% of whom were female). Core subjects include marine fishing, ocean engineering, telecommunications, aquaculture, and marine food products (Fisheries Yearbook Editorial Committee 2009).

At the university level, the National Fisheries University and the Tokyo University of Marine Science and Technology are the two national institutions of higher education specializing in fisheries science. Other national and public universities with fishery faculties or departments include Hokkaido University, Tohoku University, the University of Tokyo, Mie University, Kyoto University, Fukui Prefectural University, Hiroshima University, Kôchi University, Kyushu University, Nagasaki University, and Kagoshima University. Private universities with similar programs include Kitasato University, Nihon University, Tokai University, and Kinki University. These universities are also given financial support from public funds. Altogether, fisheries education is offered at more than 18 institutes of higher learning.

As an example of the higher fisheries education in Japan, the following is a list of departments at the National Fisheries University.

1. Department of Fisheries Information and Management: this department develops knowledge in the areas of industry management, distribution, systems, and the state of domestic and international fishing industries and teaches information processing technology. The syllabus fosters specialists who can work in the management and administration fields for the development of the fishing industry.
2. Department of Fishery Science and Technology: this department provides specialized knowledge and technical skills in navigation and the sustainable and scientific production of marine resources. The syllabus fosters specialists who can take leadership roles in the development and improvement of fishery production technology and work in environmental conservation and resource management.
3. Department of Ocean Mechanical Engineering: this department provides specialized knowledge and technical skills in marine engines and equipment, environmental sensors and equipment, and marine machinery. The syllabus fosters specialists who can work in fields related to ocean environmental conservation and the sustainable use of marine resources.
4. Department of Food Science and Technology: this department provides a grounding in the physiology and biochemistry of hygiene management, physical characteristics, and the health benefits of marine produce. Students also gain an understanding of advanced applied technology for little-used and unused resources. The syllabus fosters specialists who can work in fields related to the development and supply of safe and functionally superior fishery food products.

5. Department of Biological Production: this department provides specialized knowledge on the biological functions, breeding, and maturation habitats of fisheries plants and animals and develops techniques that apply this knowledge to resource cultivation. The syllabus fosters specialists who can work in fields related to the breeding and cultivation of resources and the protection of coastal fishing grounds.

1.4.2.2 Statistics System

The fisheries statistics system is used to gain an understanding of the current status and dynamics of fishery production and employment as well as to analyze the comprehensive industrial structures associated with fisheries, such as fishing villages, distribution, and processing. These statistics are key resources for planning fisheries policy and obtaining future perspectives on the industry. The following publications are the primary fisheries statistics published by the MAFF.

1. Fishery Census (every 5 years): this is conducted concurrently nationwide, targeting all families and businesses engaged in fisheries as well as fishery management organizations, fishery product distributors, refrigeration plants, and fish processing plants. The census includes sea fishery surveys (a survey of fishery management units, of fishery employee households, of fishery management organizations, and of sea fishery regions), inland waters fishery surveys (a survey of inland waters fishery management units and of inland waters fishery regions), and distribution and processing surveys (a survey of fishery product distributors and of refrigeration plants and fish processing plants).
2. Annual Statistics of Fishery and Aquaculture Production: this includes type of fishing, scale of operations, waters in which operations take place, production volumes by species, production volumes by type of fishing, number of fishing days, and production totals.
3. Annual Report on Fisheries Business Survey: this includes the scale of business management, income breakdowns, expenditure breakdowns, profits and losses, assets, and labor figures.
4. Annual Fishery Product Distribution Statistics: this includes distribution volumes, distribution costs, and destinations (product types) for landing-district fishery products, consumer-district fishery products, refrigerated products, and processed products.
5. Statistical Tables on Fishing Cooperatives: this includes organizational outlines of cooperative associations, their activities, and their financial status.
6. Annual Statistics on Fisheries Labor Survey: this includes the number of fishery operators, breakdowns by age, ratio of self-employed workers to employed workers, breakdowns by gender, and the number of days worked.
7. Fishing Vessel Statistics: this includes total tonnage, breakdown by horsepower, breakdown by hull material, and type of fishery.
8. Survey Report on Recreational Fishing Volumes (every 5 years): this includes estimates of catch volumes by fish variety caught by recreational fishers.

9. Food Balance Sheet: this is prepared following the United Nations Food and Agriculture Organization Guidelines.
10. Fishery Trade Statistics: this includes trading partners in fishery products, import and export items, import and export volumes, value of imports and exports, and currency exchange rates.

Individual prefectures and municipalities also gather and publicize more detailed information on local fisheries.

1.5 Research and Assistance

Another important soft infrastructure is the national system of fisheries research and its assistance to and education of local fishers.

1.5.1 Research and Technical Development

Unlike most research conducted at universities, government-run fisheries research and technical development institutes supply rapid technical solutions to pressing issues facing fisheries and also provide practical expertise in response to various fishery policies and industry needs. Government-run fisheries research and technical development in Japan can be categorized into the following types.

The Fisheries Research Agency (FRA) is Japan's national-level research body. The FRA maintains 9 institutes and 16 research centers around the country. Research areas within the FRA include economic and policy analyses of the industry, resource assessments, development of aquaculture technologies, development of applied processing technologies, research and monitoring of ocean and fishing-ground environments, marine engineering research, development of incubation and hatching technologies for aquaculture and fish farming, development of untapped resources and fishing grounds, and genetic research.

Prefectural fisheries research stations are set up at the local government level and conduct research and technical development tailored to local ecosystem conditions and social needs. Some typical research themes at these experimental stations include the development of unused coastal resources and resource management, the monitoring and prediction of local fish stocks (resource distribution) and ocean conditions (water temperatures, tides, winds, etc.), finding more effective fishing gear and methods that are best suited to local fishing conditions, development of technology for fish seeds and feed production, prevention and treatment of fish diseases, and monitoring and prevention of pollution in coastal waters.

One of the most successful outcomes of research at these prefectural research stations is surimi, a technology for producing fish sausages and other fish-paste products. Surimi technology is said to have been discovered by accident at a fisheries research station in Hokkaido, but today the technology has spread worldwide.

Some examples of research that have won the Award of the National Association of Fisheries Research Stations in recent years include research into breeding flounder using chromosome manipulation, technology for estimating the biomass of Walleye pollock resources by means of ordinary fish finders, technology for breeding Weather loach, development of a vaccine to treat cold-water disease in ayu (sweetfish), and technology for predicting ocean conditions.

The Agriculture, Forestry and Fisheries Research Council, established within the Ministry of Agriculture, Forestry and Fisheries, directs the course of fisheries research and technical development run by the central government. The seven members of the Council include prominent researchers as well as representatives from industry and public interest groups. This council draws up core research plans related to agriculture, forestry, and fisheries, assists research projects, and publishes research outcomes.

1.5.2 Providing Local Fishers with Knowledge and Techniques

Once new knowledge and technologies have been acquired through research and technical development, it is important to convey this knowledge and technology to local fishers in an understandable fashion. Also important is to ensure that the knowledge and techniques are actually beneficial for developing local fisheries by improving the capabilities of fishery operators. Therefore, Extension Officers are stationed at main fisheries sites in each prefecture.

Of all the fisheries administration personnel, these Extension Officers have the most frequent contact with fishers and receive the most feedback. Thus, these officers play a vital role in clearly communicating the needs and problems faced by local fishers to prefectural authorities and to the central government. In other words, they act as a liaison between the administrative and research organizations and local fishers. The Research and Technological Guidance Division in the Fisheries Agency of MAFF is responsible for training and the coordination of these activities. Below are some of the activities of Extension Officers in Hokkaido:

1. Promotion of fish farming technology: the Extension Officer provides guidance and instructions on technology for sustainable fishing operations. One example is the Pacific herring, an important fish on the West coast of Hokkaido, whose population today is stagnating. To restore this resource, the Extension Officers instruct fishers on technology for fish seeds release and resource management of the Pacific herring.
2. Education of future fishers: the officer holds educational events for elementary school and junior high school students on the attractiveness of the industry, the technology used in fish processing, the importance of resource management, and the crucial need to preserve the environment.
3. Guidance on efficient management: the Extension Officer holds classes where fishers learn techniques for streamlining the management of fishing operations, including keeping account books, ways to judge fishing business conditions, their merits, and how to solve problems.

4. Promotion of technology to add more value to fishery products: the Extension Officer provides instruction and suggestions on ways to manufacture processed products that use local fish as well as lessons on how to cook it. Examples include sea urchin processing classes for the wives of fishers, held in the southern part of Hokkaido, which is noted for its sea urchins. The Extension Officer provides guidance in these classes on how to treat and process sea urchin to add value to sea urchin products.
5. Guidance on how to expand sales channels: the Extension Officer gives guidance on how to promote local marine products and various sales promotion techniques, and runs events to promote exchanges between fishery operators and consumers.

Another system for promotion is on the local fishers' side. The prefectural governors officially appreciate fishers who take leadership roles in local fishers' organizations as "Fishery Mentors (gyogyô-shi)." Fishery Mentors are fishers of good character, with insight and management skills, who take a leading role in the education of young fishers. There are three categories of Fishery Mentors: Senior Fishery Mentors, Junior Fishery Mentors (under 45), and Women Fishery Mentors.

They act as counterparts, on the fishing village side, to the Extension Officers for promotion and instruction activities. They attend training sessions and meetings at prefectural fisheries research stations, actively acquire new technology and knowledge, and convey this knowledge, alongside the Extension Officers, to local fishers and provide them with guidance and instructions.

References

Asada Y, Hirasawa Y, Nagasaki F (1973) Fishery management in Japan (FAO Fisheries Technical Paper 238). Food and Agriculture Organization of the United Nations, Rome

Central Intelligence Agency (2010) The CIA world factbook 2010. Skyhorse Publishing, Washington, DC

Economic and Social Research Institute of Cabinet Office (2009) National accounts for fiscal year 2007. Media Land Co., Ltd., Tokyo

Fisheries Agency (2009) White paper on fisheries. Nourin-Toukei-Kyokai, Tokyo (in Japanese)

Fisheries Yearbook Editorial Committee (2009) Fisheries yearbook of 2009. Suisansya, Tokyo

Food and Agricultural Organization of the United Nation (2004–2008). FAO food balance sheet. Food and Agricultural Organization of the United Nations, Rome

Food and Agricultural Organization of the United Nations (1993) Report of the FAO/Japan expert consultation on the development of community-based coastal fishery management systems for Asia and the Pacific (FAO Fisheries Report 474). Food and Agricultural Organization of the United Nations, Rome

Kaneda Y (1995) Fisheries and fishing methods of Japan. Seizan-do, Tokyo

Kaneko R, Ishikawa A, Ishii F, Sasai T, Iwasawa M, Mita F, Moriizumi R (2008) Population projection for Japan: 2006–2055 outline of results, methods, and assumptions. Jpn J Popul 6(1):76–114 (in Japanese)

Makino M, Matsuda H (2011) Ecosystem-based management in the Asia-Pacific area. In: Ommar R, Perry I, Cury P, Cochrane K (eds) Coping with global changes in social-ecological systems. Wiley-Blackwells, London, pp 322–333

Ministry of Agriculture, Forestry and Fisheries (2009) Food balance sheet of 2008. Nourin-Toukei-Kyokai, Tokyo (in Japanese)

Ministry of Internal Affairs and Communications (2009) Annual report on the family income and expenditure survey. Ministry of Internal Affairs and Communications, Tokyo (in Japanese)

Nagasaki F (1994) Beef-eating culture and fish-eating culture. Nousangyoson-bunka-kyokai, Tokyo (in Japanese)

Ocean Policy Research Foundation (2004) White paper on ocean. Ocean Policy Research Foundation, Tokyo (in Japanese)

United Nations (2007) World population prospects: the 2006 revision. (Working Paper No. ESA/P/WP.202). http://www.un.org/esa/population/publications/wpp2006/wpp2006.htm. Accessed 30 June 2010

Chapter 2
A Brief Institutional History of Japanese Fisheries Management

Abstract Since the first legal provision for fisheries operations in the eight century, the fundamental concept of fisheries management in Japan has been fisheries management by the resource users themselves. This concept has been passed down to even the most recent management system, such as total allowable catches (TACs) which was adopted in Japan in 1997, and the Resource Recovery Plan that was established in 2001. In the history of institutional development, there were two big events: the modernization period (Westernization after the feudal era) in the late 1860s, and the post-WWII period while under U.S. occupation in the late 1940s. The legislative processes of major laws in these two periods are presented in this chapter, which concludes with a summary of the institutional features of Japanese fisheries management.

2.1 The Pre-feudal Era

The northwestern part of the Pacific Ocean is blessed with very high marine productivity, and the people living on the Japanese archipelago have exploited these marine resources for thousands of years. There are more than 2,500 shell mounds along the Japanese coast. More than 100 species of finfish, 300 species of shellfish, the bones of marine mammals, and fishing gear such as stone spears and bone fishhooks have been excavated from these shell mounds. They demonstrate that permanent fishing and shellfish-gathering settlements were established at least by 10,000 BC–300 BC (Ruddle 1987). Archaeological fieldwork has revealed the typical species harvested in this period to be sea bream, sea bass, bastard halibut, tuna, yellowtail, Spanish mackerel, etc., for marine fish; salmon, catfish, carp, dace, chub, eel, etc., for freshwater fish; and short-neck clam, hard clam, bloody cockle, oyster, surf clam, etc., for shellfish (Nagasaki 1994).

Around the sixth century, Japan entered the Iron Age, and people began to use iron spears and fishhooks. It is thought that primitive beach seine fishing and net

M. Makino, *Fisheries Management in Japan: Its institutional features and case studies*, Fish & Fisheries Series 34, DOI 10.1007/978-94-007-1777-0_2,

fishing also started around this period (Ministry of Education, Culture, Sports, Science and Technology 2009). There are a lot of Tanka (Japanese-style poems) about fishermen and fishing in this period. For example, below are two famous poems compiled in the oldest Japanese anthology, the Man'yoshu (*The Ten Thousand Leaves*) edited around the late seventh and eighth centuries.

Poem by Prince Kadobe (?–745) on looking out from Naniwa and seeing the lights from the fishermen's lampfires

Gazing out,
I see the points of fires
  that fishermen have lit
    in Akashi Bay,
  like the yearning for my wife
    that has flared from me. (Vol. 3-326) (quoted from Levy 1981)

Poem by Prince Takayasu (?–742) on sending a maiden a gift of wrapped silver carp

I have rowed out to the offing
  and walked along the shore,
to catch for my woman
the little silver carp
  that hide among the seaweed. (Vol. 4-625) (quoted from Levy 1981)

The first legal provision relating to fisheries operations is found in the compilation of Taiho Code from 701 AD, set up to build a centralized government following the administrative framework of the Tang dynasty (618–907) of China. In the Taiho Code, resources in the mountains, rivers, scrubland, bogs, and coastal areas were basically open to all, and not subject to taxation. This shows a clear contrast to agricultural land at the time, for which specific land users were identified and levies were imposed by the central government. In other words, marine areas were for common use and were managed by local users themselves. This basic concept was passed down to and adopted by successive rulers (Table 2.1).

After the aristocratic regime, which lasted until the Heian Era (794–1192), the first Samurai (warrior class) government was established by the Minamoto Dynasty. This is called the Kamakura Era (1192–1333). In this period, the economy was land based. Feudal landlords appointed by the central government independently governed their own land. In 1232, the government set up the first legal code legislated by the warrior class, called the Goseibai-shikimoku, which was imposed on all feudal lands. Fisheries management strictly followed the Taiho Code and stated that resources in mountains, rivers, scrubland, bogs, and coasts were basically for common use.

After a long period of unstable government and conflict, the Warring States Period, Japan entered a stable feudal era, the Edo Era (1603–1868) governed by the Tokugawa Dynasty. The resultant social stability led to an increase in population. Cities such as Edo (then Tokyo) and Osaka were among the largest cities in the world at that time: according to studies on historical demography, for example, in the early eighteenth century, the total population of Edo exceeded one million, making it a bigger city than London or Paris at the time. This growth in population increased demand for seafood, and led to the development of coastal fisheries.

Table 2.1 Changes in fisheries institutions in Japan

Period	Area	Institutional frameworks
Up to the Edo era (–1603)	The marine areas were for common use, and managed by local users themselves	
Early Edo era (1603 – about 1700)	Coastal areas	Communities controlled the areas, and were responsible for establishing appropriate rules governing the use of these areas
	Offshore	Basically open access. Anyone could operate, regardless of the location of the home community
Later Edo era (about 1700–1868)	Coastal areas	Development of labor-intensive and capitalized fisheries. A few wealthy fishermen monopolized fishing operations
	Offshore	Large-scale fisheries operators established their own guilds and made rules, protected by feudal lords
Modernization period (1868–1901)	The government tried to introduce a top-down fishery management system, but the scheme failed. There was a return to the customary arrangement in which local fishermen controlled and managed local fishing operations	
Meiji Fishery Law (1901–1949)	Coastal areas	Fishing rights, as exclusive real rights, were granted to both Fisheries Societies (i.e., local fishermen's organizations) and individuals
	Offshore	Fishing licenses were issued to individuals or juridical persons
Present Fishery Law (1949–)	Coastal area	Fishing rights, as limited real rights, are granted to both Fisheries Cooperative Associations (i.e., local fishermen's organizations) and individuals
	Offshore	Fishing licenses are issued to individuals or juridical persons

Modified from Makino and Matsuda (2005)

In this period, each village was a formal administrative unit, and taxes and levies were charged per village according, largely, to the population and production within the village territory. The head of each village functioned as a government official, and among other duties, he kept the domiciliary registrations of villagers, collected rice-based taxes charged to his village, and applied levies to villagers such as forced labor.

In 1743, the Tokugawa Dynasty introduced a set of standard fishery regulations. This standard was called the "Urahô" or beach law, and relevant regulations in each feudal domain were supposed to be based on this law (Hirasawa et al. 1992). Its basic concepts were (1) coastal fishing grounds in nearshore waters should be used only by the people from local fishing communities; and (2) offshore fishing grounds should be left open for free access by any fishers.

According to this standard, coastal waters were regarded as extensions of the land and thus a part of the feudal domain (Akimichi and Ruddle 1984; Caddy and Cochrane 2001). The feudal lords partitioned the coastal waters and allocated them to local communities under the control of the village heads. In general, communities controlled adjacent coastal areas, and were responsible for establishing appropriate rules for use of the area. Under these rules, qualified individuals living in the community were entitled to engage in fishing activities.

The role of the community, effectively an autonomous management body of local fishers, constituted the basis for subsequent Fisheries Societies (Sect. 2.2.2), as well as for present-day Fisheries Cooperative Associations (FCAs; Sect. 2.3.3). Offshore fisheries, however, were basically open access; anyone could fish there, regardless of the location of the fishermen's home community. The boundary between coastal and offshore waters was mainly determined by depth. For example, the length of a paddle was used as a standard.

Around the middle of the Edo Era, the population further increased and technological progresses enabled the development of labor-intensive, capitalized fisheries such as beach seine fisheries, large set-net fisheries, seaweed aquaculture, and whaling. As a natural result, a few wealthy fishing people monopolized coastal fishing operations. Large-scale offshore fisheries operators established their own guilds and made their own rules. These regimes were valued and protected by feudal lords in exchange for contributions, and they functioned as formal institutions. According to trial records of the time, law courts applied offshore guild rules even to nonmembers of the guild (Makino 2003).

2.2 The Modernization of Japan

2.2.1 The Meiji Revolution of 1868

The arrival of Commodore Perry from the United States in 1853 brought an end to the national seclusion policy that had lasted for more than 200 years. The Tokugawa Dynasty was finally overthrown in 1868, and the new government (the Meiji government) carried out a radical modernization of the whole national institutional framework. At this time, for the Japanese government, modernization meant Westernization, and many laws and social institutions were replaced by European-style versions. For example, the constitutional monarchy with its Emperor followed the English model, the administrative system followed that of Germany, civil law was imported from France, etc.

As for fisheries management, the central government abolished the old feudal institutional arrangements. Then, in 1875, nationalization of Japan's seas and a centralized license system was adopted. It constituted a system of rent for the usage of nationalized sea areas. In a top-down approach, the central government tried to control and manage fishing operations through the issuance of licenses. In this system, each fisher maximized his benefit subject to private costs, which included the license fee set by the government.

After the adoption of this drastic institutional change, many individuals, who had not previously been fishing people, successfully applied for licenses. According to the national statistics on fisheries production, annual fisheries production tripled in 7 years. This phenomenon suggests several conclusions. First, the large number of new fishing people indicates that, apart from distributive equity, feudal era fisheries organization had been lucrative only in terms of rent. Second, we can conclude that the observed dramatic growth in fishery production meant that fishing pressure on the environment was effectively controlled during the feudal era. Clearly, such a rapid increase in fishing efforts could easily create a situation in which the overexploitation of fishery resources could occur.

Presumably as a result of overexploitation, the sharp rise in fish production following this institutional change turned out to be a temporary phenomenon, and total catch levels soon dropped. Widespread conflict subsequently developed among fishing people. To deal with these problems, the Bureau of Fisheries was established within the Ministry of Agriculture and Commerce in 1885. In addition, the government enacted the Fishermen's Union Regulation in 1886, by which fishing people were encouraged to establish local Fishermen's Unions. This regulation was the first formal recognition that fishing people's organizations could operate as local management authorities. According to the 1886 regulations, the fishing peoples' unions should eliminate disputes and confusion through mutual consultation among the fishing people themselves. In summary, the imposition of top-down fishery management had failed, and the institutional framework returned to its original arrangement, in which local fishing people controlled and managed their fishing operations themselves.

Another major item of legislation in this period was the Ocean Fisheries Promotion Law of 1897, in which financial support was provided from the government to fishing people operating in offshore areas. Offshore fisheries rapidly developed, spurred by technological innovations such as automated fishnet knitting machines in 1888 and engine-powered fishing vessels in 1906. Typical offshore fisheries types in this period were trawl fisheries, purse seine fisheries, and Norway-style whaling.

2.2.2 *The Meiji Fisheries Law*

In 1901, the Meiji Fisheries Law was enacted (and amended in 1910). This law put fishery rights and licenses, for the first time, into statutory form. Fisheries licenses were issued to individuals or juridical persons for offshore and distant water fisheries. Fishery rights were granted to both Fisheries Societies (local fishing people's organizations) and individuals, and were classified into four categories: (1) set-net fishery rights; (2) specific fishery rights for beach seines, boat seines, etc.; (3) aquaculture rights for oysters, pearls, etc.; and (4) exclusive fishery rights (Yamamoto 1995). Exclusive fishery rights were further classified into traditional exclusive fishery rights (which could be granted to an individual, based on customary use in the feudal era), and new exclusive fishery rights (newly granted to local Fisheries Societies by the central government). These exclusive fishery rights were area-based rights and included all the resources present in or migrating through the area.

The nature of these fishery rights was, in effect, that of property rights. Although the expiration period was fixed by law, revision of rights was virtually unconditional. Especially after the 1910 amendment, fishery rights became exclusive real rights that could be sold, leased, transferred, or collateralized.

Theoretically, the use of an area's resources or fisheries operations should be continuously revised according to technological changes and other socioeconomic or environmental changes. However, the nature of exclusive property rights and unconditional revisions resulted in fragmented ownership of fishing grounds and also allowed rights holders to arbitrarily exercise exclusivity. Through the processes of free collateralization and transferability, fishery rights became concentrated in the hands of just a few people, who effectively controlled coastal areas in the same way as feudal landlords had in the past. Many fishing people without fishery rights were exploited by absentee rights owners, merchants, or middlemen; the exploited fishing people worked like serfs.

In sum, the defects of the Meiji Fisheries Law were as follows: the exclusivity and unconditional revision of area-based fishery rights prevented the flexible and adaptive coordination of resource use, and free transfer and collateralization led to the monopolization of coastal fisheries. In addition, there were no coordinating mechanisms to link the various fisheries operations within and among areas, and the exclusive and absolute exercise of legal rights prevailed.

2.3 Current Fishery Laws

2.3.1 Fisheries Reform After WWII

The end of WWII and the Allied Occupation, from August 1945, brought dramatic and sweeping institutional changes to Japan, including the adoption of the current constitution. Following agrarian land reform, the General Headquarters of Supreme Commander for the Allied Powers (GHQ) requested the reform of fisheries institutions in a democratic manner, and the current Fisheries Law was enacted in 1949.

The legislative process of the current fishery law could be viewed as a kind of political game played between capitalism (GHQ: the USA) and socialism (the then Soviet Union): a game that included rent-seeking processes by vested fishing people and small-scale fishing people. Planning of a new Fisheries Law (i.e., the current Fisheries Law) started in June 1946, and the law was enacted in November 1949. Major events and modifications of the original bills are summarized in Fig. 2.1.

In the first bill, all fishery rights were to be granted only to fishing people's organizations, but vested rights holders strongly opposed this. The proposed system seemed more socialist than capitalist or market-based and, on this basis, the Soviet Union supported it but GHQ rejected it. For the third bill, GHQ issued instructions to establish a personal rights system with free transfer, free collateralization, and semipermanent duration, i.e., a market-based rights system based on

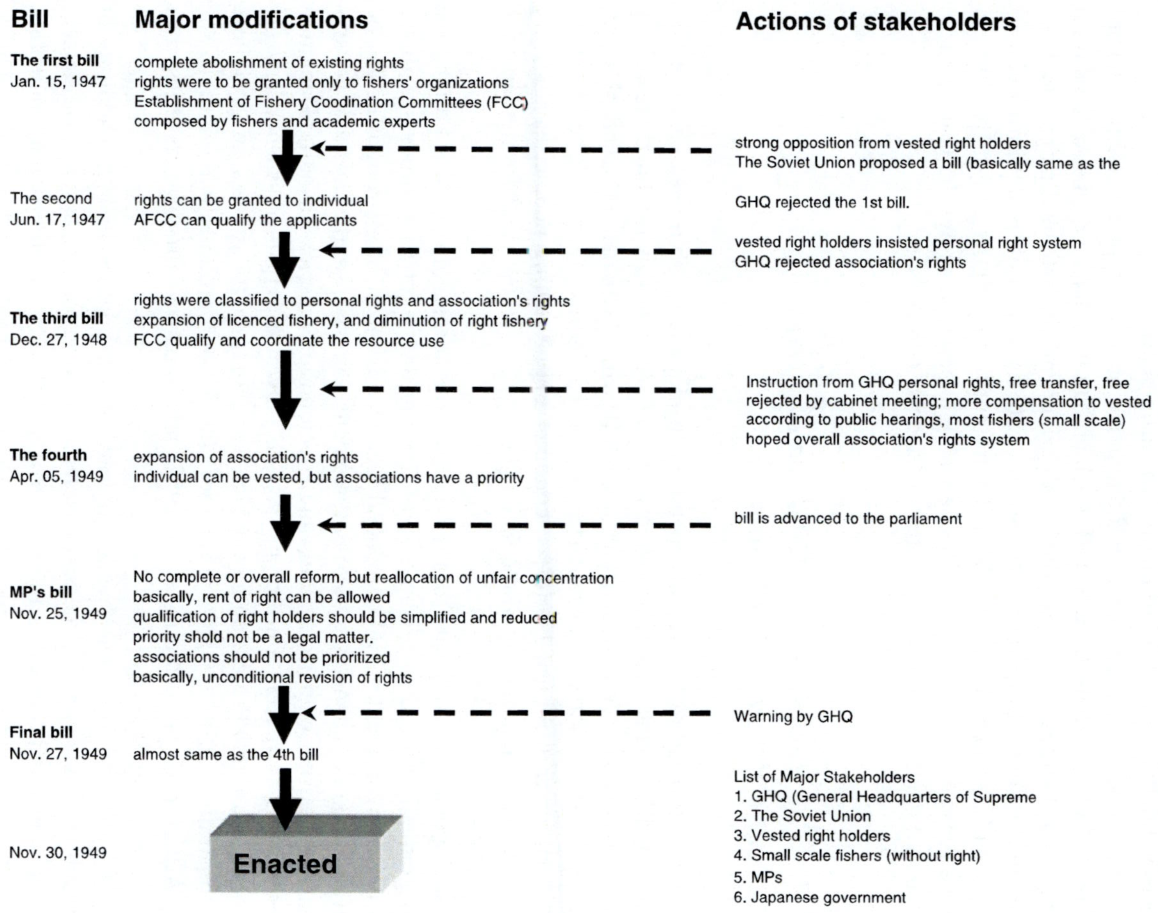

Fig. 2.1 Legislative processes leading up to the current Fishery Law of 1949 (Modified from Makino 2001)

property rights. However, the Japanese government recognized that this kind of system had constituted the fundamental drawback of the Meiji Fisheries Law, as described in the last section, and did not comply. It took the Japanese government more than 2 years to persuade GHQ of the importance of establishing fishing people's organizations and the inadequacy of personal property rights for Japanese fisheries; finally, the fourth bill made it to parliament.

A crucial point was reached as parliament members attempted to change the bill in favor of vested rights holders. GHQ, which had initially given instructions for a market-based rights system, warned the parliament against the changes, and the final bill (the fourth bill) was made into law, with fishery rights that prioritized fishing people's organizations (details of the current fishery rights and license system will be provided in Sect. 2.3.3). This Japanese experience shows the potential role of external powers in drastic institutional reforms such as these.

2.3.2 *The Resumption of Offshore Fisheries and Resource Protection*

Right after the end of the war, all operations by Japanese fishing vessels were limited to small coastal areas designated by GHQ. The lines that defined the fishing grounds were called "the MacArthur Line," following the name of the U.S. General. Because of food shortages in Japan, as well as the increased population of fishing communities with returned soldiers, conflicts among fishing people became very severe, making enlargement of the fishing grounds an urgent issue for the Japanese government.

In response to a request from the Japanese government, the MacArthur Line was gradually moved to offshore areas. During the negotiation process with the Japanese government, GHQ demanded the establishment of a strong official system for resource management and enforcement. In response, the Japanese government set up fisheries patrol vessels and enacted the Fisheries Resource Depletion Prevention Law of 1950. Further, in 1951, this law was expanded as a new law, the Fisheries Resource Protection Law, which is still in force today. As a result of these administrative efforts, GHQ's limitation on fishing grounds according to the MacArthur Line was abolished in 1952, and Japanese offshore fisheries then showed very rapid growth.

The Fisheries Resource Protection Law of 1951 states that the Minister of Agriculture, Forestry and Fisheries or Prefectural Governors can establish ordinances or regulations on fisheries operations. Also, waters suitable for spawning, nurseries, feeding, etc., are designated by the Minister as Protected Waters to protect aquatic animals and plants. In Protected Waters, the governor of the Prefecture makes plans for increasing aquatic animals and plants. Any construction or land reclamation is strongly restricted in this area.

Special measures are provided in this law for anadromous species like salmon, which migrate from rivers to the sea and back again. The law prohibits catching salmon in inland waters, including rivers and lakes. Also, to protect the routes taken by anadromous fish, the owners of constructions such as dams are required to install fish ladders so as not to interrupt upstream fish migration.

2.3.3 Fisheries Management Systems

2.3.3.1 Fishery Rights and Licenses

Under the current Fisheries Law of 1949, marine fisheries are classified into three categories: (1) fishery rights for coastal fisheries, (2) fishery licenses for offshore and distant water fisheries, and (3) other fisheries. Coastal fishery rights are classified, in turn, into (1a) Common Fishery Rights (only for Fisheries Cooperative Associations (FCAs)), (1b) Large-scale Set-net Fishery Rights, and (1c) Demarcated (aquaculture) Rights (Asada et al. 1973; Shima 1983; Nagasaki and Chikuni 1989; Yamamoto and Short 1992; FAO 1993; Kaneda 1995).

Fishery rights are granted by the prefecture governors based on the "Fishery Ground Plan" drafted by representatives of the local fishing people (to be detailed below). Although the expiration period is fixed in law, fishing rights are regarded as real rights, and the provisions of the territorial rights law are applied mutatis mutandis. However, they do not include the right to privatize the sea surface into portions. Fishing rights are somewhat similar to use rights in their attributions, i.e., the right to conduct fishery operations exclusively in specified areas by specified methods. Legally, fishery rights cannot be sold, leased, transferred, or collateralized. Fishery licenses, on the other hand, are not real rights, but taking the large capital investments of the license holders into account, they are also strongly protected in the practical sense.

1a. The Common Fishery Right is the right to engage in "Common Fishery," which means the common use of resources within specified coastal areas by specified local fishing people. There are five types of Common Fishery. Type 1 Common Fisheries are operations targeting seaweed, kelp, shellfish, or other sedentary species in the area. Type 2 are fisheries using small-scale fixed nets such as small-scale set-net fisheries or gillnet fisheries. Type 3 includes beach seine fishing, trawl fishing using non-powered vessels, etc. Type 4 includes wintering mullet fishing, etc. Type 5 is operations in rivers and lakes, commonly called Inland Water Fishery Rights. Common Fishery Rights are granted only to the local FCAs, and the FCAs are mandated to establish their own autonomous regulations for sustainable use by their members.

1b. The Large-Scale Set-Net Fishery Right is the right to operate fisheries using set-nets over 27 m in depth. This right can be granted to individuals. However, since large-scale set-nets monopolize good coastal fishing grounds for extended periods, the local FCAs or other organizations of local fishing people are prioritized. The purpose of this priority is to allow as many local fishing people as possible to participate in the operation of this fishery.

1c. The Demarcated (aquaculture) Fishery Right is the right to engage in aquaculture. There are three types: The Type 1 Demarcated Fishery involves installing facilities in a certain area, e.g., stakes or sticks for seaweed, oysters, pearl oysters, etc., hanging rafts for oysters, pearls, scallops, etc., and floating net cages for yellowtail, sea bream, etc. Type 2 is large-scale aquaculture in an area surrounded by barrier facilities such as net cages. Type 3 is other aquaculture operations such as seabed cultivation of scallops, short-neck clams, etc. More details of the current aquaculture operations are presented in Sect. 3.3.

2. Fishery licenses lift a prohibition in a specific case for generally prohibited fishing, but this action does not create any rights to be legally protected. However, as mentioned above, they are protected in a practical sense. They are broadly divided into two types: Minister-licensed fisheries (called Designated Fisheries) and Governor-licensed fisheries. Almost all the industrialized large-scale fisheries operating offshore and in distant waters are categorized as Minister-licensed fisheries. These fisheries are directly managed by the minister of MAFF (Ministry of Agriculture, Forestry and Fisheries). Governor-licensed fisheries are medium-scale fisheries operating around the coast and offshore areas. The current situation of licensed fisheries is presented in Sect. 3.2.

2.3.3.2 Nested System for Fisheries Coordination

In the immediate aftermath of WWII, food shortages were an urgent national issue. According to government documents made available in 1963, the principal aim of fisheries reform at this time was to develop fisheries' productivity to cope with the domestic food shortage, and to improve the economic status of fishing people actually engaged in fishing operations. To achieve this goal, the overall objective of the government at that time was the efficient and extensive development of fishery resources without overexploitation. The government recognized a strong need for the enhancement and conservation of fishery animals and plants as a prerequisite of the reform.

How could this goal be achieved? How could resources be utilized more efficiently, extensively, and sustainably? The methodology was the "holistic utilization of sea surfaces". This is the most distinctive feature of the current Japanese institutional framework, as explained below.

In contrast to agricultural land, sea areas can be utilized in three dimensions. At a fishing ground, a wide range of target species can be harvested using various kinds of equipment. In addition, a fishing operation conducted by one person can, by its nature, influence others' operations, either actually (physically) or potentially. Few fisheries, especially finfish fisheries, can be conducted without using a certain minimum amount of sea area. It is therefore necessary to arrange and coordinate various fishing operations within a certain area from an overall point of view, and not simply from the viewpoint of each economic unit. This is termed "holistic fisheries coordination." Various levels and scales of coordinating organization have been instituted to facilitate holistic fisheries coordination (Table 2.2).

Table 2.2 Coordinating organizations in Japan

Level	Organizations	Functions
National	Fishery Policy Council	The advisory body to the government for national level fishery coordination, design of national fishery policy, international issues, etc.
Multi-jurisdictional	Wide-Area Fisheries Coordinating Committees (WFCCs)	Coordination of resource use and management of highly migratory species. Also addresses resource recovery plans
Prefectural	Area Fisheries Coordinating Committees (AFFCs)	Mainly composed of elected fishers. Coordination through the Fishery Ground Plan, Prefectural Fishery Coordinating Regulations, and Committee Directions
Local	Fisheries Cooperative Associations (FCAs)	Composed of local fishers. They establish operational regulations (FCA regulations) that stipulate gear restrictions, seasonal/area closures of fishing grounds, etc.
More specialized	Fishery Management Organizations (FMOs)	Autonomous body of fishers. FMO rules are more detailed and stricter than the FCA regulations

The smallest-scale coordinating organizations are local Fisheries Cooperative Associations (FCAs). There are 1,041 FCAs in 2008 (MAFF 2010). To achieve holistic fisheries coordination for local fishing grounds, local FCAs have to establish operational regulations (FCA regulations) that stipulate equipment restrictions, as well as closures of fishing grounds on a seasonal or area basis, etc.

Area Fisheries Coordinating Committees (AFCCs) are established in each prefecture. Basically, each AFCC consists of nine elected fishing people, four academic experts, and two representatives of the public interest (usually local government officials). The AFCC's considerable power and authority is made explicit in the Fisheries Law. All rights (based on the Fishery Ground Plan) and the governor licenses (based on the Prefectural Fishery Coordinating Regulations) are granted by prefectural governors, following recommendations or advice from the AFCC. In effect, the AFCC decides the allocation of fishing rights and licenses in areas within their jurisdiction. The AFCC can also restrict the attributions of fishing rights and licenses, and can issue Committee Directions as appropriate. The objective of Committee Directions must be to promote the enhancement and conservation of fishery animals and plants to achieve efficient and extensive fishery production, without violating sustainability. The AFCC can request a Prefectural Governor's Order to enforce compliance, on the part of fishing people, and give directions.

Wide-Area Fisheries Coordinating Committees (WFCCs) were established by an amendment to the Fisheries Law in 2001. There are three WFCCs in Japan: Pacific, Japan Sea and Western Kyusyu, and the Seto Inland Sea. They are chiefly composed of elected committee members from each AFCC, and act at levels higher than that of prefectural jurisdiction. WFCCs coordinate resource use and management of highly migratory species, and address Resource Recovery Plans for widely distributed or migrating species, as explained in Sect. 2.4.2.

The highest-level coordinating organization is that of the Fishery Policy Council. This council constitutes the advisory body to the government with respect to national-level fisheries coordination, design of national fishery policy, etc.

In addition to these formal coordinating organizations, a number of operational ideas have been developed since the late 1970s, largely on the initiative of local fishing people. These developments include what is known as Shigen Kanri-gata Gyogyo, literally meaning "Resource Management-type Fishery" (Hasegawa 1989; Baba 1996; Nakanishi 2005). To maintain and improve incomes, as well as sustain resources, various management measures have been initiated by autonomous bodies of fishing people, called Fishery Management Organizations (FMOs). In 2008, there were 1,738 FMOs in Japan (MAFF 2010), and they take a wide range of forms. They often constitute a group of fishing people within an FCA. Sometimes, FMOs comprise members from several neighboring FCAs or even from FCAs of several prefectures. Fishers using the same equipment or targeting the same species also form FMOs. Cases of Shigen Kanri-gata Gyogyo and the role of FMOs are presented in Chap. 4.

To sum up the current fisheries coordinating system, the Fisheries Law simply provides a framework for management via a system of fishery rights and licenses. To achieve holistic utilization of sea surfaces, coordinating organizations with wide-ranging authority and power have emerged. For example, AFCCs can decide how to allocate, and restrict the application of, fishing rights/licenses using the Fishery Ground Plan and Committee Directions. In addition, a variety of fishing restrictions have been stipulated in prefectural fishery coordinating regulations, FCA regulations, and FMO rules. Prefectural Fishery Coordinating Regulations broadly stipulate fishing restrictions, in order that the regulations may be applicable throughout the prefecture. FCA regulations stipulate more detailed fishing restrictions that are relevant to local conditions. These FCA regulations take into account the restrictions set out in the Prefectural Fishery Coordinating Regulation but may also include restrictions that have not been stipulated in the prefectural regulations. In the same manner, the FMO rules are even more detailed and yet stricter than the FCA regulations.

2.4 Recent Systems for Fisheries Management

2.4.1 The Total Allowable Catch (TAC) System in Japan

In July 1996, Japan ratified the United Nations Convention on the Law of Sea (UNCLOS). In accordance with the ratification, the Government of Japan enacted the Law Regarding Preservation and Management of Living Marine Resources. With this law, the Total Allowable Catch (TAC) system was introduced in 1997.

Table 2.3 Total allowable catches in 2008

Species	Total allowable catch (in 1,000 ton)	Real amount of catch
Pacific saury	455	346.3
Walleye pollock	239	207.9
Japanese jack mackerel	271	154.0
Japanese sardine	52	31.1
Chub mackerel and spotted chub mackerel	765	471.2
Japanese common squid	333	196.9
Snow crab	7.8	4.8

As explained above, the traditional approach of formal fisheries management in Japan takes the form of input control and technical control. The government controls the total amount and quality of fishing pressure via rights and licenses that specify gear, fishing season, size of vessel, target species, etc. This makes the TAC system the very first output control measures to be adopted in Japan.

The TAC system directly manages the catch of specified species with the upper limit of total catch in tonnage. At present, seven TACs apply to eight species: Pacific saury (*Cololabis saira*), walleye pollock (*Theragra chalcogramma*), Japanese jack mackerel (*Trachurus japonicus*), Japanese sardine (*Sardinops melanostictus*), chub mackerel (*Scomber japonicus*) and spotted chub mackerel (*Scomber australasicus*), Japanese common squid (*Todarodes pacificus*), and snow crab (*Chionoecetes opilio*) (Fisheries Agency and Fisheries Research Agency 2009).

Every year, the Fisheries Research Agency and prefectural research stations conduct research ship surveys of the seas around Japan, and assess the status of fish stocks. In the Stock Assessment Reports, the Fisheries Research Agency also makes recommendations on the total catch amount (called Allowable Biological Catch [ABC]). Then, taking account of the results of Stock Assessment Reports and socio-economic conditions of the fishery type, the Fishery Policy Council discusses the values of TACs for eight species, and the Minister of Agriculture, Forestry and Fisheries decides the final TACs. Each TAC is divided into two categories: the minister-managed TAC, and the prefectural governor-managed TAC. Both TACs are then allocated to fisheries sectors managed by the minister (offshore industrial fisheries managed by licenses from the minister) and prefectural governors (coastal small- and medium-scale fisheries managed by rights and licenses from the governors). Table 2.3 shows the values for TACs in 2008. Taking operational and market conditions into account, the annual catch amounts are totaled from April to March for walleye pollock and snow crab, from July to June for chub mackerel and spotted chub mackerel, and from January to December for other species.

In major fisheries sectors such as offshore bottom trawlers, large- and medium-scale purse seiners, longliners, etc., a committee for TAC enforcement is organized within the organization of vessel owners. After receiving a portion of TAC allocated to the sector by the national government, the committee plans how to allocate the TAC among areas and seasons. In other words, the fishing people's organizations are responsible for determining the access rules. For example, the TAC committee in the Pacific saury fishing people's organization makes a harvest plan every 10 days. The committee also holds the chief responsibility for announcing its TAC and ensuring that its members comply with it. For example, the TAC committee of the northern Pacific purse seiners autonomously set the individual vessel quotas for chub mackerel in 2008 to make sure the total catch did not exceed the allocated amount (Sect. 5.2).

2.4.2 *Resource Recovery Plan*

In 2001, the Resource Recovery Plan (RRP) system, a new resource management framework, came into force. This new system aims to strengthen the basic approach of the Resource Management-type Fishery (Shigen Kanri-gata Gyogyo) with official support from the government.

As shown in Sect. 2.3.1, a Resource Management-type Fishery is based on the initiative of local fishing people, and measures are implemented by Fishery Management Organizations (FMOs, Table 2.2) on an autonomous basis. They implement various management measures such as the minimum catch size limit, enlargement of mesh sizes of fishing nets, adoption of no-take periods or no-take areas, adoption of individual catch quotas, and artificial fish seed release.

However, because of its autonomous nature, a Resource Management-type Fishery has at least three major weaknesses. First, the majority of the management activities are implemented by the FMO composed of fishing people using the same equipment within one FCA. Therefore, it is difficult to deal with cross-jurisdictional issues for migrating species management, or conflicts among users of different types of fishing gear. The second weakness is that, because the decision-making process in most of the FMOs depends on complete unanimity, it is very difficult to implement innovative or drastic measures, and it is sometimes impossible to solve the problem. Finally, the main aim of the measures implemented by FMOs is to increase their income, and objective evidence of sustainable resource use is generally weak. The RRP system was established to deal with above three weaknesses.

Several official measures were newly established under the RRP system. First, with an amendment of the Fisheries Law, a new coordinating organization for multi-jurisdictional issues, named the Wide-Area Fisheries Coordinating Committees (WFCCs, Table. 2.2) was established in 2001. There are three WFCCs in Japan: the Pacific WFCC, the Seto Inland Sea WFCC, and the Japan Sea and western area WFCC. These WFCCs have the authority to address the RRPs on wide-distributing or migratory species, as well as the Committee Directions on their enforcement.

Another new measure, established in 2001, is the Economic Support System. To encourage depleted species to recover, drastic and innovative measures are often required. Also, for species that need several years for maturation, these measures for recovery must last for years. Additionally, in many cases where a resource has become significantly depleted, the economic conditions of local fishing people have also worsened, and they are not able to agree to the adoption of drastic measures even if they are willing. Therefore, the Economic Support System compensates a proportion (two thirds or a half) of the economic losses brought about by drastic or innovative measures under the plan, and promotes consensus-building among fishing people.

The third measure for RRPs is the Total Allowable Efforts (TAE) system established in 2001 under an amendment of the Law Regarding the Preservation and Management of Living Marine Resources. If some fishermen suspend operations or cut their number of vessels to allow resources to recover, if other fishing people operate more intensively, or if new fishing people start operations on that species, the recovery will ultimately fail. TAEs are the upper limits on fishing efforts to catch specified species to avoid situations such as this. TAEs are defined as total number of operating days.

The drafting processes of the RRP are as follows. First, the Area Fisheries Coordinating Committees (AFCCs in Table 2.2) survey and discuss the status of various fisheries resources in their jurisdictional area. If the fish are widely distributed or migratory species, the Wide-Area Fisheries Coordinating Committees (WFCCs) provide the main arena for this process. If the AFCC or WFCC decides that resource recovery measures should be implemented for specific species or fishery types, a Fishers' Council for the Recovery Plan, consisting of representative fishing people harvesting that species, is organized. The Fishers' Council, government officers, and researchers then jointly discuss the contents of the RRP. Researchers provide key scientific information on the recovery scenario, and set the numerical goal of the RRP (target level and time frame) so as to ensure accountability for sustainable use of the resource.

In 2010, a total of 66 RRPs were implemented (Fisheries Agency 2010). In Sect. 5.2, a case of chub mackerel RRP from the north Pacific purse seiners is presented. TAEs are applied to nine species: Pacific sandlance (*Ammodytes personatus*) in the Sôya Channel, Japanese Spanish mackerel (*Scomberomorus niphonius*) in the Seto Inland Sea, Pacific false halibut (*Hippoglossoides dubius*) in the western Japan Sea, the tiger puffer (*Takifugu rubripes*) in Ise and Mikawa Bays, roughscale sole (*Clidoderma asperrimum*) in the northern Pacific, brown sole (*Pleuronectes herzensteini*) in the northern Japan Sea, marbled sole (*Pleuronectes yokohamae*) in the western Seto Inland Sea, willowy flounder (*Tanakius kitaharai*) in the northern Pacific, and spear squid (*Loligo bleekeri*) in the southern Pacific. Table 2.4 shows their TAEs in 2008–2009.

2.4.3 Other Major Laws and Systems

The Marine Fisheries Resource Development Promotion Law of 1971 was amended in 1990, and the "Resource Management Agreement System" was established. This system encourages autonomous agreements among fishing people for the purpose of resource management. If an agreement prevails at a certain level within the area, the government can affirm the agreement and make it an official rule.

Table 2.4 Total allowable efforts in 2008–2009

Species	Fishery types	TAE (total days)
Pacific sandlance	Offshore bottom trawl	616
Japanese Spanish mackerel	Driftnet	100,324
	Hanatsugi-ami (small-scale purse seine)	2,020
	Middle-scale purse seine	1,288
	Trawl	74
Pacific false halibut	Offshore bottom trawl (single vessel)	6,210
	Offshore bottom trawl (paired vessel)	575
	Small-scale bottom trawl	6,450
Tiger puffer	Small-scale bottom trawl	7,953
Roughscale sole	Offshore bottom trawl	6,243
	Small-scale bottom trawl	616
Brown sole	Offshore bottom trawl	729
	Small-scale bottom trawl (Type 1)[a]	2,521
	Small-scale bottom trawl (other types)	1,843
	Gillnet	5,246
Marbled sole	Small-scale bottom trawl	16,260
Willowy flounder	Offshore bottom trawl	6,565
	Small-scale bottom trawl	3,696
Spear squid	Offshore bottom trawl	342

[a]Small-scale bottom trawling (Type 1) is a bottom trawl fishing operation using powered vessels of less than 15 gross tons without net mouth-spreading devices

It constitutes another official support system for autonomous resource management by fishing people.

One of the latest cases of application of the Resource Management Agreement System is chub mackerel and Japanese jack mackerel fisheries in Oita Prefecture, Kyushu Island. These two species, harvested in the Bungo Channel (between Kyushu and Shikoku Islands), are highly valued and branded in the Japanese market. They are harvested using two methods: pole-and-line fishing (647 coastal fishing people) and purse seiners (6 offshore fleets). These two fisheries have been in intense conflict for more than 20 years. After seeing the research results obtained by the Oita Prefectural Fisheries Research Station, both fishery types agreed to add no-take days during the spawning season (May and June) to protect spawning stocks. This agreement became a Resource Management Agreement in June 2010. It may be the starting point for closer coordination on this shared resource by the two fishing types. The signing ceremony was attended by the leaders of both fisheries organizations and the Governor of Oita Prefecture.

In 2001, the "Basic Act on Fisheries Policy" was enacted to deal with the changes in circumstances surrounding Japanese fisheries, such as the decreasing self-sufficiency or advancing age of workers in the fisheries sector (to be detailed in Chap. 3). This law aims to establish a symbiosis between producers and consumers, and between cities and fishing communities, by establishing a new policy framework for Japanese fisheries in the twenty-first century (Ono 2002). There are two basic principles in this act (Articles 2 and 3): (1) securing a stable supply of fisheries products, and (2) the healthy development of fisheries sectors. The government will formulate a "Basic Plan for the Fisheries Policy" to set out the basic principles, in which the target for the self-sufficiency rate in fishery products will be included. The latest Basic Plan was formulated in 2007, and will be reviewed, in principle, every 5 years.

In 2007, the "Basic Act on Ocean Policy" was enacted to clarify the principles of the ocean policy in Japan, covering all the marine-related industries, including fisheries. The six basic principles set down in this act are: (1) harmonization of the development and use of the oceans in ways that conserve the marine environment, (2) securing safety and security on the oceans, (3) improvement of scientific knowledge of the oceans, (4) sound development of ocean industries, (5) comprehensive governance of the oceans, and (6) international partnership with regard to the oceans. Based on this act, the Headquarters for Ocean Policy, headed by the Prime Minister of Japan, was established in the Cabinet, and the Basic Plan on Ocean Policy was formulated in 2008. This Basic Plan prescribes in more detail the direction of ocean policy in Japan.

2.5 Discussion

It has often been said that Japanese fisheries institutions are well established; this characteristic has been attributed to their historical development during the feudal era. In reality, however, it has been a process of trial and error. A top-down system was adopted at the beginning of the Meiji era and, after it dissolved into chaos, there was a return to local organization-based institutions. The Meiji Fisheries Law could not respond to technological changes or the increasing control of rights by powerful entities. Fisheries reform processes after WWII came perilously close to failure.

The basic concept of marine areas being for common use and managed by local users themselves has been implemented, but the nature of fishing rights and licenses has substantially changed. Under the current institution, the exercise of full fishery rights and licenses is restrained by the inherent legal requirements for resource conservation; and various coordinating organizations of fishing people, including autonomous regulations and agreements, play vital roles in this process. Government or research agencies also play an important role in fishery management and resource management. They provide support to fishing people in the form of administrative advice or scientific information. However, local fishing people remain the principal decision makers. Resource management rules set by FCAs or FMOs can be tailored to fit local environments; these rules are generally flexible and responsive to specific needs. These autonomous resource management systems are suitable, therefore, for adaptive resource management. To sum up, the authorities and responsibilities of fisheries management are shared with fishing people and governments, and can be categorized as a form of co-management (Makino and Matsuda 2005). Cases in Chaps. 4 and 5 will show examples of how this system works.

Several lessons may be learned from the Japanese experience. First, as illustrated by what happened at the beginning of the Japanese modernization process, sudden changes or replacements of institutional frameworks are not always successful. Complete failure can result. Second, unconditional personalization of fishery rights or sea areas can lead to overconcentration of fishing rents and inflexible use of resources. Some form of coordinating organization, whether governmental or nongovernmental, is indispensable. Third, in some instances, external powers may play a vital role in the process of resource rent redistribution. Finally, the roles of government or public research stations continue to be important in co-management regimes, even after co-management frameworks are well established. They can compensate the weakness in autonomous measures via legal, financial, and scientific supports.

Several emerging issues in the Japanese institutional framework merit recognition and discussion. The first of these concerns fishery coordination by fishing people. Although coordination of this type is vital for the current institutional framework, it inevitably becomes very complex and locally specific. Sometimes fishing people cannot play their expected roles as coordinators or as members of AFCCs. In addition, some fishing people or coordinating organizations have proved to be unwilling to adopt new technologies, and have held back technical progress. Successfully managed areas may have particularly strong tendencies in this direction. Second, although the number of fishing people in Japan continues to decline, the number of recreational fishermen and pleasure-boat owners is growing. These recreational users are now important stakeholders. However, they are not fully included in the current coordinating systems, and their position is very weak compared to that of professional fishermen. Third, there is the question what justifies fishery rights. Fishers use marine resources that are the common property of all citizens, but the benefits they derive from fishing are protected by law. Fisheries operations that lack sustainability and responsibility cannot be justified over the long run, even if local stakeholders support them. Therefore, measures pertaining to the legal responsibilities that come with resource and ecosystem stewardship need to be included in any statement of fishery rights.

References

Akimichi T, Ruddle K (1984) The historical development of territorial rights and fishery regulations in the Western Pacific. In: Ruddle K, Akimichi T (eds) Maritime institutions in the Western Pacific. National Museum of Ethnology, Osaka, pp 37–88

Asada Y, Hirasawa Y, Nagasaki F (1973) Fishery management in Japan (FAO Fisheries Technical Paper 238). Food and Agriculture Organization of the United Nations, Rome

Baba O (1996) The present status and issues in the resource management-type fisheries. J Reg Fish 37(1):39–49 (in Japanese)

Caddy JF, Cochrane KL (2001) A review of fisheries management past and present and some future perspectives for the third millennium. Ocean Coast Manage 44(9–10):653–682

Fisheries Agency (2010) Resource recovery plan. http://www.jfa.maff.go.jp/j/suisin/s_keikaku/index.html. Accessed 30 June 2010 (in Japanese)

Fisheries Agency and Fisheries Research Agency (2009) Assessment of the fishery resources at the waters surrounding Japan. http://abchan.job.affrc.go.jp/digests21/index.html. Accessed 30 June 2010 (in Japanese)

Food and Agricultural Organization of the United Nations (1993) Report of the FAO/Japan expert consultation on the development of community-based coastal fishery management systems for Asia and the Pacific (FAO Fisheries Report 474). Food and Agricultural Organization of the United Nations, Rome

Hasegawa A (1989) Theory and types of the resource management-type fishery. Jpn J Fish Econ 33(2–3):1–39 (in Japanese)

Hirasawa Y, Sato M, Yamamoto T (1992) Fisheries cooperatives as the core for development of coastal fisheries management system- Japanese experience. In: Yamamoto T, Short K (eds) International perspectives on fisheries management. National Federation of Fisheries Cooperative Associations, Tokyo, pp 127–158

Kaneda Y (1995) Fisheries and fishing methods of Japan. Seizan-do, Tokyo

Levy IH (1981) The ten thousand leaves – a translation of the Man'yoshu, Japan's premier anthology of classical poetry. Princeton University Press, Princeton

Makino M (2001) The legislative history of the fisheries reform after the WWII. Soc Syst Res 4:61–75 (in Japanese)

Makino M (2003) The institutional history and international features in Japanese fisheries management. Nippon Suisan Gakkaishi 69:368–375 (in Japanese)

Makino M, Matsuda H (2005) Co-management in Japanese coastal fishery: it's institutional features and transaction cost. Mar Policy 29:441–450

Ministry of Agriculture, Forestry and Fisheries (2010) Fisheries census of 2008. Ministry of Agriculture, Forestry and Fisheries, Tokyo (in Japanese)

Ministry of Educational, Culture, Sports, Science and Technology (2009) Fisheries. Jikkyo Syuppan, Tokyo (in Japanese)

Nagasaki F (1994) Beef-eating culture and fish-eating culture. Nousangyoson-bunka-kyokai, Tokyo (in Japanese)

Nagasaki F, Chikuni S (1989) Management of multispecies resources and multi-gear fisheries (FAO Fisheries Technical Paper 305). Food and Agricultural Organization of the United Nations, Rome

Nakanishi T (2005) Resource management-type fisheries. In: Japanese Society of Fisheries Economics (ed) The productions and foresight in the fisheries economics. Seizando-syoten, Tokyo, pp 87–92 (In Japanese)

Ono S (2002) Basic act on fisheries policy and fisheries researches. Nippon Suisan Gakkaishi 68:242–246 (in Japanese)

Ruddle K (1987) Administration and conflict management in Japanese coastal fisheries (FAO Fisheries Technical Paper 273). Food and Agriculture Organization of the United Nations, Rome

Shima K (1983) The role of cooperatives on the exploitation and management of coastal resources in Japan (FAO Fisheries Technical Paper 296). Food and Agricultural Organization of the United Nations, Rome

Yamamoto T (1995) Development of a community-based fishery management system in Japan. Mar Resour Econ 10:21–34

Yamamoto T, Short K (1992) International perspectives on fisheries management. National Federation of Fisheries Cooperative Associations (Zengyoren), Tokyo

Chapter 3
Japanese Fisheries Today

Abstract This chapter describes, using the latest statistics, current conditions and problems facing Japanese capture fisheries, including total production, business conditions, resource levels, trade, self-sufficiency rate, etc. Next, profiles of major capture fisheries sectors and their management frameworks are described. Aquaculture (mariculture) is another important sector, especially in production values in coastal areas. The history, institutional frameworks, and major species cultured are described. Fish ranching, which involves the release of artificially produced fish seeds into natural waters, is very active in Japan. Current activities and problems facing fish ranching activities are described. Finally, the social roles and economic size of the fisheries processing sector, which is also a large fisheries sector in Japan, are briefly summarized.

3.1 Japanese Fisheries: Current Conditions

3.1.1 Production

Japan is one of the world's largest fishing nations with respect to both production and consumption. In 2006, fisheries production (both capture and aquaculture) was 5.6 million tons. It was ranked the fifth in the world, following China, Peru, the USA, and Indonesia (FAO 2008a). Another international feature of Japanese fisheries is the variety of species harvested. Table 3.1 shows the species composition of Japanese fisheries production in 2008. According to the latest five-year average of capture fisheries production by FAO FISHSTAT, 33 species account for 90% of the total catch volume (4.4 million tons) in Japan. A similar country in terms of the variety in catch composition is Spain, in which 46 species constitute 90% of the total of 0.9 million tons. The other end of the spectrum is the high-latitude northern European countries. For example, in Norway (2.7 million tons), 90% of the total catch is

M. Makino, *Fisheries Management in Japan: Its institutional features and case studies*,
Fish & Fisheries Series 34, DOI 10.1007/978-94-007-1777-0_3,

Table 3.1 Species composition of Japanese fisheries production in 2008 (in 1,000 ton)

Species	1,000 ton
Marine capture fisheries: total	4,367.5
Fish: total	3,358.5
Tuna	216.9
Bluefin tuna	20.5
Southern bluefin tuna	1.6
Albacore tuna	50.2
Big eye tuna	65.8
Yellowfin tuna	77.3
Others	1.4
Striped marlin	3.2
Swordfish	9.5
Indo-Pacific blue marlin	4.4
Other marlin	1.4
Skipjack tuna	303.6
Frigate tuna and bullet tuna	27.9
Shark	32.9
Salmon	180.4
Gizzard shad	7.4
Pacific herring	3.5
Japanese sardine	37.7
Round herring	48.0
Japanese anchovy	344.8
Japanese jack mackerel	172.1
Chub mackerel and spotted chub mackerel	514.0
Pacific saury	354.5
Japanese amberjack (yellowtail)	76.1
Bastard halibut	7.3
Flounder	55.6
Pacific cod	42.2
Walleye pollock	212.2
Arabesque greenling	169.8
Channel rockfish	1.5
Sandfish	15.0
Deep-sea smelt	4.9
Conger eel	6.3
Swordfish	16.2
Japanese red sea bream	15.8
Chicken grunt	4.9
Japanese Spanish mackerel	15.7
Japanese sea bass	10.5
Sand eel	62.0
Tilefish	1.4
Puffer fish	5.2
Other fish	373.6

(continued)

Table 3.1 (continued)

Species	1,000 ton
Crustaceans: total	22.4
Japanese spiny lobster	1.4
Kuruma prawn	0.7
Snow crab	5.3
Red queen crab	20.2
Swimming crab	2.8
Krill	41.6
Others	25.7
Shellfish: total	401.6
Abalone	1.7
Spiny top shell	8.1
Short-neck clam	39.1
Scallops	310.2
Others	42.5
Squid: total	291.4
Japanese common squid	216.1
Neon flying squid	26.2
Other squid	49.2
Octopus	48.3
Sea urchin	10.8
Marine mammals	1.3
Other marine animals	53.1
Sea plants: total	104.7
Kelp	73.3
Other sea plants	31.4
Marine aquaculture: total	1147.9
Fish: total	262.7
Coho salmon	12.8
Japanese amberjack and greater amberjack (yellowtail)	158.3
Japanese jack mackerel	1.7
White trevally	2.7
Japanese red sea bream	71.0
Bastard halibut	4.2
Puffer fish	4.1
Other fish	7.9
Shellfish: total	417.3
Scallop	225.6
Oyster (with shell)	190.4
Other shellfish	1.3
Kuruma prawn	1.6
Tunicates	10.7
Other marine animals	0.2
Sea plants	455.4
Kelp (konbu)	46.8
Sea mustard (wakame)	54.6

(continued)

Table 3.1 (continued)

Species	1,000 ton
Seaweed (nori)	337.9
Mozuku	15.8
Other sea plants	0.3
Pearls (ton)	2.5
Inland capture fisheries: total	32.7
Salmon	10.5
Pond smelt	1.1
Ayu	3.5
Icefish	0.4
Carp	1.4
Eel	0.3
Goby	0.3
Other fish	3.1
Freshwater clam	9.8
Other shellfish	1.3
Prawns	0.8
Other marine animals	0.4
Inland aquaculture: total	39.9
Rainbow trout	6.8
Other trout	3.1
Ayu	5.9
Carp	3.0
Eel	21.0
Other fish	0.2

Source: MAFF (2009a)

composed of 8 species: blue whiting, Atlantic herring, Atlantic cod, coley, capelin, brown seaweed, Atlantic mackerel, and haddock. For Iceland (1.8 million tons), it is seven species: capelin, blue whiting, Atlantic herring, Atlantic cod, haddock, coley, and golden redfish, and for Sweden (0.3 million tons) it is only five species: Atlantic herring, European sprat, sandlance, blue whiting, and Atlantic cod.

Figure 3.1 shows the summary of changes in Japanese fisheries productions from FY 1960 to 2005. As detailed in Sect. 1.2, there are several ways of classifying fisheries sectors in Japan. In this figure, Japanese fisheries are divided into five sectors: coastal fisheries (up to 10 gross tons), offshore fisheries, distant-water (high sea) fisheries, marine aquaculture, and inland aquaculture. In the late 1960s and early 1970s, distant-water fishing was the most important fisheries sector in Japan. However, after the establishment of the 200-nautical-mile regime worldwide, the importance of this type of fishery drastically declined. On the other hand, in the late 1970s and the 1980s, offshore fisheries developed. The total volume peaked in 1984, producing 12.8 million tons. The main catch of this period in volume was the Japanese sardine (*Sardinops melanostictus*). However, because of natural fluctuations in biomass (Baumgartner et al. 1992), landing of Japanese sardines declined severely in the early 1990s (Watanabe et al. 1995; Yatsu et al. 2005).

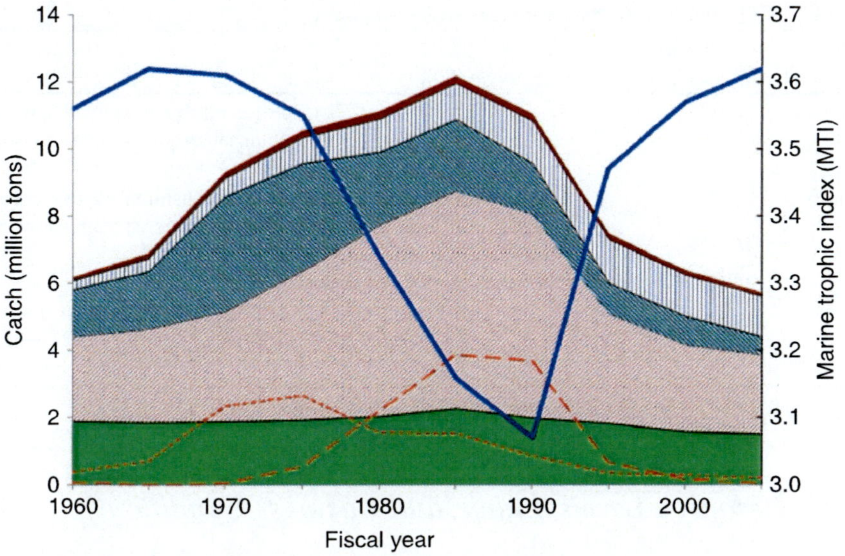

Fig. 3.1 Total production of Japanese fisheries and the Marine Trophic Index (MTI) from 1960 to 2005, divided into five categories: coastal fisheries, offshore fisheries, distant-water fisheries, marine aquaculture, and inland aquaculture, from bottom to top. Broken and dotted lines represent catches of sardine and walleye pollock, respectively (Modified from Matsuda et al. 2010)

On the other hand, coastal fisheries have shown relatively stable production since the 1960s, with a slight decline over the last 20 years. They have occupied the most important sector in production values since the late 1980s. At the same time, marine aquaculture developed considerably, and now it is the second largest sector in terms of production value.

It is worth discussing the changes in catch structure. Many scientists warn of the degradation of global marine ecosystems (Myers and Worm 2003; Worm et al. 2006). In this respect, Pauly and Watson (2005) proposed an index based on the mean trophic level of fisheries catches, called the Marine Trophic Index (MTI), and showed that the MTI of the global fisheries has decreased from ca. 3.5 in 1950 to ca. 3.3 in 1990. This implies overfishing, since the harvested fish are increasingly coming from the less valuable lower trophic levels as populations of higher trophic level species are depleted. Pauly et al. (1998) called this trend "fishing down." Based on these studies, the Convention on Biological Diversity has selected the MTI as an indicator of marine ecosystem integrity and ecosystem goods and services in global biodiversity outlook.

The theory of "fishing down" is based on the assumption that the major target species is high priced and sits at a higher trophic level. Nevertheless, this is not always true (Delgado et al. 2003). Also, the MTI of the global marine landings did not show a monotonic decline but fluctuated from decade to decade. The global MTI was low in the 1970s and 1980s, when catches of Peruvian anchovy and Japanese sardine, respectively, were high. The bold line in Fig. 3.1 shows the Japanese MTI from 1960 to 2005 (Matsuda et al. 2010). The MTI was ca. 3.6 in 1960, less than 3.1 in 1990, and ca. 3.6 in 2000, so did not show monotonic decline.

Table 3.2 Stock levels of fishery resources around Japan

Stock level	Number of stocks	Examples
High	14	Saury (north Pacific stock), spotted chub mackerel (Pacific and East China Sea stocks), Japanese common squid (autumn stock), etc.
Medium	28	Japanese jack mackerel (Pacific and Tsushima warm current stocks), Japanese common squid (winter stock), snow crab (northern Pacific and Japan Sea stock), etc.
Low	42	Chub mackerel (Pacific and Tsushima warm current stocks), Japanese sardine (Pacific and Tsushima warm current stocks), walleye pollock (northern Japan Sea and Pacific stocks), etc.

Source: Fisheries Agency and Fisheries Research Agency (2009)

3.1.2 Resource Levels, Labor, and Business Conditions

According to a recent stock assessment report, about the half of the 84 stocks officially assessed are at low levels (Table 3.2). Various factors are behind this decline in resources, including changes in the marine environment such as regime shifts (Kawasaki 1983), loss of seagrass beds and tidelands for the spawning and growth of fish through coastal land reclaims or modification, and overfishing.

In Japan, the total areas of seagrass beds and tidelands have been steadily decreasing over the last several decades. For example, in 1945, there were 82,621 ha of tidelands along the Japanese coastline. In 1996 they had an area of 49,380 ha, corresponding to an approximately 40% decrease in the last 50 years (Environment Agency and Marine Parcs Center of Japan 1994). These decreases are particularly noticeable in densely populated coastal areas such as Tokyo Bay, Osaka Bay, and the Seto Inland Sea. Cases of seagrass bed restoration activities initiated by local people are described in Sect. 7.3.

The number of people engaged in fisheries production in 2007 was 204,000. This figure has been continuously decreasing since 1953, when the figure was about 800,000. The advancing age of fishers is also a serious problem. Currently, 48% of male fishers are over 60 years old (Table 3.3). According to one report, 63.7% of fishers have no one to take over their business, while only 19.5% have someone in mind (Nôrin-Chûkin Research Institute 2008). Because the number of new fishermen is very small (about 1,200–1,500 persons per year), the decrease in the total number and increase in the average age will continue in the near future.

The total number of fishing vessels registered in 2007 was 313,000, of which 9,000 were non-powered and 267,000 were of less than 5 gross tons (Fisheries Agency 2009). Therefore, compared to other developed countries such as Iceland, New Zealand, Canada, and the USA, most of Japanese fishers are small-scale operators in coastal areas (Makino and Matsuda 2011).

The business conditions of the average Japanese fishers are not favorable. Conditions in two classes of capture fisheries are shown below based on the statistical

Table 3.3 Composition of male fishermen in 2007

Age group	Percentage in total male fishers (%)
15–24 years old	2.7
25–39 years old	11.5
40–59 years old	37.9
60–64 years old	10.5
Over 65 years old	37.4

Source: MAFF (2009b)

Table 3.4 Business conditions in the average coastal fisher household in 2007

Item	Amount (in JPY 1,000)
Fishery revenue	6,716
Fishery expenses	
– Labor costs	441
– Equipment costs	335
– Maintenance costs	252
– Fuel costs	821
– Commission fees	417
– Depreciation	575
– Others	1,133
Other revenue	122
Total income for household	2,864

Source: MAFF (2009c)

definitions of the Ministry of Agriculture, Forestry and Fisheries (MAFF). The average income of coastal capture fishers' households (households with fishers using vessels of less than 10 gross tons) was JPY 2.86 million in 2007. This figure is much lower than the national average of all employees of JPY 5.56 million. Table 3.4 shows the composition of their revenues and expenses. The largest expense is that of fuel (20.7%), so increases in fuel prices have a serious effect on their businesses.

The business conditions endured by medium- and large-scale fishers (industrial fishers) are much worse. In 2008, the weighted-average profit of industrial fishers using vessels of over 10 gross tons was JPY 18.8 million, not counting depreciation costs, and JPY –3.7 million with depreciation costs included. This means that typical fishers in this class will not be able to replace their vessels in the foreseeable future.

3.1.3 Trade and Self-sufficiency Rate

Japan is one of the largest consumers of seafood in the world. In 2007, Japan imported 2,892,000 ton of fisheries products, costing JPY 1,637 billion. This was the world's greatest in value, and the second in volume following China (Ministry of Finance 2009).

The four major imported products by value are JPY 214 billion of prawns from Vietnam, Indonesia, India, etc., JPY 197 billion of tuna from Taiwan, Korea, Australia, China, etc., JPY 130 billion of salmon from Chile, Russia, Norway, etc., and JPY 70 billion of crab from Russia, Canada, etc. The biggest three exporters to Japan by country are China (JPY 334 billion for eel, processed crab, tuna, etc.,) the USA (JPY 141 billion for cod, cod roe, sablefish, etc.,) and Russia (JPY 120 billion for crab, cod roe, salmon, etc.).

Japan exported about one million tons of fisheries products in the late 1980s, worth more than JPY 300 billion. The major export items in this period were fish meal (sardines) and canned seafood. However, due to the sharp decline in sardine resources in the late 1980s, Japanese exports shrank to 612,000 ton, worth JPY 238 billion, in 2007. This amounted to 0.3% by value of total exports from Japan.

As for export items, pearls have been the most important for several decades, amounting to JPY 37 billion in 2007. The next most important export item is dried sea cucumber, worth JPY 17 billion. With economic development and increased demand in China, exports of dried sea cucumber to Hong Kong have dramatically increased since around 1995, and sea cucumber resources around the Japanese coast have suffered heavy fishing pressure (a case of sea cucumber fisheries management is presented in Sect. 4.2). The third export item was tuna (JPY 15 billion). The most important destination for Japanese exports is Hong Kong (JPY 64 billion). The most important item in value is pearls, while exports of dried products such as sea cucumber, abalone, and scallop eyes have been steadily growing in recent years. Other important destinations are the USA (JPY 36 billion for scallop, pearl, yellowtail, etc.) and Korea (JPY 33 billion for walleye pollock, sea bream, mackerel, etc.).

The self-sufficiency rate is one of the most important policy issues in Japan. In 1964, the self-sufficiency rate of fisheries products was 113%, but it has since steadily decreased to about 60%. In 2007, the self-sufficiency rate for fisheries products was 62%. Based on the Basic Plan for the Fisheries Policy (Sect. 2.4.3), the Fisheries Agency has set the target for self-sufficiency at 65% in 2017.

3.2 Profiles of Major Fisheries Sectors

3.2.1 Distant-Water and Near-Water Tuna Fisheries

Tuna resources harvested by Japanese tuna fishers are managed under the Regional Fisheries Management Organizations (RMFOs) such as the Western and Central Pacific Fisheries Commission (WCPFC), the Indian Ocean Tuna Commission (IOTC), the Commission for the Conservation of the Southern Bluefin Tuna (CCSBT), the International Commission for the Conservation of Atlantic Tuna

(ICCAT), and the Inter-American Tropical Tuna Commission (IATTC). The ICCAT set the limit for annual bluefin tuna catches in the eastern Atlantic Ocean and Mediterranean Sea to 13,500 ton in 2010, down from 22,000 in 2009; and the quota for Japan was decreased accordingly to 1,148 ton from 1,871 ton. The WCPFC adopted measures to decrease fishing pressure on bigeye tuna, such as moratoriums, no-take zones, and cuts in total catch amounts. CCSBT also cut the annual catch limit of southern bluefin tuna from 11,810 ton in 2009 to 9,449 in 2010.

Because Japan is one of the biggest consumers of tuna, all fisheries operations by Japanese vessels, as well as tuna consumption by the Japanese, need to be managed according to the rules set by regional fisheries management organizations (RFMOs). To ensure this, the Japanese government amended the ministerial ordinance on tuna fisheries in 2008. Individual vessel quotas were imposed on Atlantic bluefin tuna fishing. Four government officers were newly allocated as Tuna Resource Inspectors and tasked with inspecting imported frozen tuna. The overall inspection system was also tightened.

The private sector also plays an important role in tuna fisheries management. For example, the Organization for the Promotion of Responsible Tuna Fisheries (OPRT) is an international nongovernmental organization based in Tokyo. It comprises tuna longline producers in Japan, Taiwan, Korea, the Philippines, Indonesia, China, Ecuador, the Seychelles and Fiji, as well as traders, distributors, and consumer organizations. They publish a Positive List of tuna fishing vessels (authorized vessels complying with rules by their flag states and RFMOs) and promote bycatch reduction technologies.

3.2.2 Medium-Scale Salmon Driftnet Fishing

Mother-ship-type salmon fisheries were developed in Japan from 1951. In 1959, 16 mother ships and 460 catcher boats were in operation. However, there are no mother ships now, and current harvests are mainly from driftnet fisheries in offshore areas and set-net fisheries in coastal areas. Japan ratified the Convention for the Conservation of Anadromous Stocks in the North Pacific Ocean, and stopped salmon fishing on the high seas in 1992. Now all salmon driftnet fishery operations are conducted within the Japanese and Russian Exclusive Economic Zones (EEZs).

Most of the salmon harvested by driftnet fishing within Japanese EEZ are thought to originate in Russian rivers, so Japan and Russia hold an annual meeting and set salmon driftnet catch quotas within the Japanese EEZ. In 2008, the catch quota of Russian-derived salmon within the Japanese EEZ was 3,005 ton, and the agreement included financial and technical support from Japan to Russia.

Since 2007, the middle-scale and small-scale driftnet fishermen (vessel size of up to 30 gross tons and licensed by the governor of the Hokkaido Prefecture) have operated within the Russian EEZ based on the conditions set by intergovernmental consultations, such as catch quotas, number of vessels, royalties, etc. In 2008, the catch quota for these two types of Japanese driftnet fishermen within the Russian EEZ was 9,735 ton from May 11 to July 31. The royalty was JPY 307 per kilogram.

3.2.3 *Distant-Water Bottom Trawl Fishing*

This fishery engages in bottom trawling in distant-water (high seas) areas with vessels of more than 15 gross tons. In the 1960s and 1970s, mother-ship-type bottom trawlers developed and landed more than one million tons with 8,000 fishers working there. However, after the establishment of the EEZ regimes, many of the fishing grounds for Japanese bottom trawlers were closed, and all the mother-ship-type trawlers were decommissioned. The remaining fishing grounds open to Japanese distant-water bottom trawlers and their operations are as follows.

In the high seas area of the Bering Strait, the walleye pollock is the main target species. Since 1995, they have been managed by the Convention on the Conservation and Management of Pollock Resources in the Central Bering Sea. Japan, USA, China, Korea, Russia, and Poland are the member countries. However, a moratorium has been imposed since 1995 because the resource status is very low.

The Emperor seamounts are a chain of submerged volcanic mountains extending northward from the northwestern Hawaiian Islands to the tip of the Aleutian Islands and Kamchatka. This area is regarded as a Vulnerable Marine Ecosystem (VME), and appropriate conservation measures need to be implemented based on the resolution adopted by the UN General Assembly in 2006 (FAO 2008b). Around these seamounts, several Japanese bottom trawlers are catching splendid alfonsino and southern boarfish. Foundation of a new RFMO is now being discussed by the governments of Japan, Korea, Russia, and the USA.

Southern blue whiting, blue grenadier, hake, and squid have been harvested by Japanese bottom trawlers around New Zealand since around 1959. In 1977, the total catch amounted to about 180,000 ton. However, in 1978, the New Zealand EEZ was set, and only two vessels are now operating there.

In the Antarctic Sea, Antarctic krill have been caught under a management framework by the Commission for the Conservation of Antarctic Marine Living Resources (CCAMLR) since 1982. Only one Japanese vessel is currently in operation. Also, in the northwestern Atlantic area, one bottom trawler is harvesting Greenland halibut under the management of the Northwest Atlantic Fisheries Organization (NAFO).

3.2.4 *Offshore Bottom Trawl Fishery*

This fishery mainly operates within the Japanese EEZ, using vessels larger than 15 gross tons. It is the second largest fishery type in Japan. Most of the 415 vessels registered in 2008 are the single-boat type, while about 10% of the total are the twin-boat type, which are allowed to operate at very limited areas. The single-boat type can then be divided into Danish seine fishing and otter trawl fishing. Danish seine fisheries operate without mouth-spreading devices.

The main target species of this fishery varies from area to area, reflecting the variety in marine ecosystems there. The size and number of vessels also vary considerably as the result of the long-year coordination between coastal fisheries in each area.

For example, in the Hokkaido area, 48 vessels of 124–160 gross tons are targeting walleye pollock, Atka mackerel, sand eel, and flounder. Some of them also operate in the Russian EEZ. In the north Pacific area, 122 vessels of 19 gross tons and 65–75 gross tons are targeting walleye pollock. Some of them are the twin-boat type. In the central and southern Pacific areas, 18 vessels of 15–40 gross tons and 125 gross tons are targeting squid, lizardfish, deep-sea smelt, etc. The north Japan Sea area has 70 vessels of about 40 gross tons, which harvest Atka mackerel, walleye pollock, shrimp, and crab. Finally, the western Japan Sea area produces sandfish, flounder, snow crab, and deep-sea smelt. At present there are 128 trawlers in operation, and a case study on snow crab fishing from this area will be presented in Sect. 5.1. The borders between trawl fishing grounds and coastal areas where trawling is totally prohibited are set based on the Fisheries Law.

3.2.5 Large- and Medium-Scale Purse Seine Fishery

This is the largest fishery sector in Japan, yielding about 20% of the total capture fisheries production. Operations are conducted using vessels over 40 gross tons (over 15 gross tons in the north Pacific area). In the 1960s, most purse seine fleets comprised two net vessels, one lighting vessel, and one or two transport vessels. About 60–80 fishers collectively operated in one fleet. However, since around the mid-1970s, the increase in oil prices triggered changes in technology, and the typical purse seiner fleet is now composed of one net vessel and one searching or lighting vessel and one or two transport vessels with about 50 fishermen. As of 2009, 207 fleets were in operation. The most productive fishing grounds are the northern Pacific, where they harvest sardine, anchovy, chub mackerel, tuna, etc. The management of chub mackerel by purse seiners in this area will be described in Sect. 5.2. In the central Pacific and Indian Ocean areas, 45 of a total of 207 fleets of purse seiners are in operation. They use larger vessels (over 200 gross tons) and target tuna all year round.

3.2.6 North Pacific Saury Fishing

Saury are harvested in two categories of fisheries: the north Pacific saury fishery, a minister-licensed fishery which uses dip net vessels of greater than 10 gross tons, and other small-scale coastal fisheries with a fisheries license obtained directly from the prefectural governor.

There were 197 vessels holding the ministerial licenses in 2009, comprising 100 vessels smaller than 20 gross tons and 74 vessels larger than 40 gross tons. The main fishing season is from August to December, and they operate salmon driftnet fishing or tuna fishing in other seasons. Because saury migrates southward from the Pacific coast of Hokkaido Island to around Chiba Prefecture (the central part of Honshu island), the saury fishers follow them, and the landing places gradually move south.

Saury resources have been managed according to the Total Allowable Catch (TAC) system since 1997. The resource status is high in recent years, and the TAC in 2009 was 455,000 ton. As explained in Sect. 3.2.4, the organization of saury fishers (both minister-licensed and governor-licensed saury fishers) comprises the TAC committee, which is composed mainly of representatives of all the saury fishers. It determines the access rules, harvest plans, and spatial allocations of the TAC. Saury is a typical commodity with high price elasticity of supply. The harvest rules and plans made by the TAC committee are thus very important for saury fishers' business stability. As mentioned above, there are two large groups of different vessel sizes within the TAC committee (less than 20 gross tons and over 40 gross tons). Very heated discussions are held between them every year when setting harvest rules and plans. Saury landing places also move more than 1,000 km, and spatial allocation is another subject for annual discussions among fishers and processors.

3.2.7 *Middle- and Large-Scale Squid Jigging Fishing*

Squid jigging fishing can be divided into three types. The first is coastal squid jigging, which uses vessels of less than 30 gross tons. Some are governor-licensed fisheries, but they vary among prefectures. These fishers operate in coastal areas and catch squid as one of their many target species. The second type is offshore jigging using a vessel from 30 to 100 gross tons, called a Middle-Scale Squid Jigging Fishery, for which fishers must hold a license from the minister. They catch squid from May to February. The third type is Large-Scale Squid Jigging Fishery using vessels over 100 gross tons. This is also a minister-licensed fishery, and they operate near New Zealand and the southwestern Atlantic areas. There are 177 vessels holding ministerial licenses.

Within the New Zealand's EEZ, the large-scale squid jigging vessels are lent to a local company, and operating based on ITQs held by the local company. In the 2007/2008 season, 2,109 ton were harvested by three vessels. In the southwestern Atlantic, several vessels are operating around Peru, Argentina, etc.

3.2.8 *Whaling*

Japan has long history and rich culture relating to whales, and this sector used to be one of the most important sectors in Japan (Akimichi 1990; Morikawa 2009).

Following the resolution on the moratorium at the International Whaling Committee (IWC), Japanese mother-ship-type whaling fleets and large-scale whaling vessels halted operations in the Antarctic Oceans in 1986 and in the Pacific in 1988 (Endo and Yamao 2007). Only nine small-scale whaling fishers are currently conducting coastal whaling, targeting Baird's beaked whales and pilot whales. These fishers must hold licenses from the minister. Dolphin fishing is also conducted in a few coastal areas based on licenses from prefectural governors.

As permitted by Article 8 of the International Convention for the Regulation of Whaling (ICRW), scientific programs with Special Permits are operated at the Antarctic Sea and in the northwestern Pacific by Japanese researchers. These programs are called JARPA (Japan's Whale Research Program under Special Permit in the Antarctic) and JARPAN (Japan's Whale Research Program under Special Permit in the Northwestern Pacific), respectively. Their aim is to collect biological data such as natural mortality rate, growth rate, cetacean roles in marine ecosystems, etc.

3.3 Aquaculture

3.3.1 History and Overview

An ancient document recounts that fish were stocked and raised for cuisine palatine in palace ponds at the end of the eighth century. The first recorded shellfish aquaculture was seabed-sown cultivation of oysters in the Seto Inland Sea in the middle of the sixteenth century. Commercial ongrowing of red sea bream began at the start of the seventeenth century. Nori (laver) cultivation using a supporting system was started by fishers living in Edo (Tokyo) at the end of the seventeenth century. In the middle of the nineteenth century, semi-intensive carp culture in rice paddy fields started, as well as eel farming in ponds.

The first intensive aquaculture of marine fish, Japanese amberjack, mackerel, and sea bream, was carried out in enclosures in 1930. Cage culture was then developed in the 1950s, leading to major gains in productivity. Until the mid-1960s, the Japanese amberjack was the most commonly cultured marine fish, but the share of red sea bream increased, and today several dozen species are cultured all over Japan. Commercial aquaculture of oysters was first developed using a support system, but this was eventually superseded by hanging culture under rafts. From the 1950s, hanging culture with long lines, characterized by its resistance to high waves, predominated, mainly in the northern region of Japan. This technique was also used to cultivate larger seaweeds such as Japanese kelp (kombu). Pearl culture first succeeded in 1893. After 1910, the production of full-orbed pearls was made possible by technical developments which have since been adopted by pearl farms worldwide (Ohshima 1994).

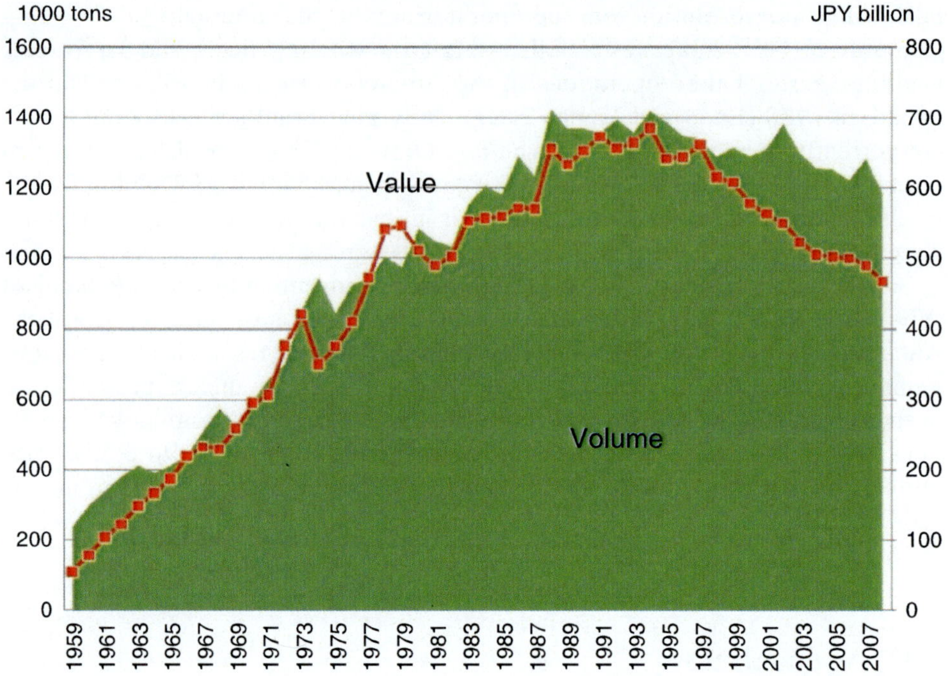

Fig. 3.2 Japanese aquaculture production for 1959–2008 (Source: MAFF, 1960–2009. Values are in real terms with 2000 as the standard year)

Currently, marine or freshwater aquaculture takes place in all 47 prefectures in Japan. According to the official statistics of the Ministry of Agriculture, Forestry and Fisheries (MAFF), estimated total aquaculture production in 2008 was 1,186 ton, corresponding to 21% of the total national fisheries production. More than 60 species are included in the aquaculture section of annual fisheries statistics. Of total aquaculture production, 96% is marine aquaculture (Table 3.1). The main cultured species are nori (seaweed), yezo scallop, oyster, Japanese amberjack, red sea bream, sea mustard, eel, sweetfish, rainbow trout, pearls, etc. Figure 3.2 shows the changes in aquaculture production in Japan. Today, aquaculture contributes to bringing previously high-priced species (such as amberjack, red sea bream, and eel) within the range of the average consumer, helping to support a varied dietary culture.

In 2008, there were 19,646 enterprises engaged in marine aquaculture, employing 96,292 workers in the high season, of whom 49,915 were women. In freshwater aquaculture, 3,764 enterprises were employing 12,494 individuals, 3,709 of whom were women. Therefore, in aggregate, the aquaculture sector in Japan supports 100,000 jobs (MAFF 2010). However, as Fig. 3.2 shows, the value of production has been continuously falling since around 1990, and the business conditions for average aquaculture operators are rapidly worsening.

3.3.2 *Legal Framework for Aquaculture*

Aquaculture inevitably requires a certain amount of sea surface to be closed off to the public. Therefore, all the aquaculture operators must have fishery rights for aquaculture (Demarcated Fishery Rights, see Sect. 2.3) as defined in Article 6 of the Fisheries Law of 1949.

The Law to Ensure Sustainable Aquaculture Production of 1999 seeks to prevent self-induced environmental deterioration around fish farms. Pursuant to this law, the MAFF issued Basic Guidelines to Ensure Sustainable Aquaculture Production and the FCAs developed and implemented "Aquaculture Ground Improvement Programs," which can be developed individually by a single FCA or jointly by more than one FCA, and which must be approved by the prefectural authorities. In addition, the Basic Environmental Law requires the government to establish Environmental Quality Standards (EQS), which are target levels for water quality that must be achieved and maintained in public waters.

As for genetically modified organisms (GMOs), in 2000, the MAFF issued Guidelines for the Application of Recombinant DNA Organisms in Agriculture, Forestry, Fisheries, the Food Industry and Other Related Industries. The purpose of these Guidelines is to establish basic requirements to govern the appropriate application of recombinant DNA organisms in agriculture, forestry, fisheries, and the food industry, as well as in other related industries regulated by the MAFF, to assure the safe use of DNA-modified organisms and achieve the sound overall development of agro industries. For more details, see FAO (2005–2010).

3.3.3 *Main Cultured Species*

3.3.3.1 Marine Species (Kumai 2005; Mori 2005)

Japanese amberjack (*Seriola quinqueradiata*) and greater amberjack (*Seriola dumerili*) are the most economically important species in Japan; they account for about 25% of the total production value of aquaculture. The Noami family in Kagawa Prefecture started enclosure aquaculture in 1930. Since the 1950s, cage culture has been widely adopted in the western regions of Japan. Today, production is around 150,000 ton, with aquaculture production double that of marine capture production. Although artificial seed production was mastered technically in the 1960s, wild seeds are still mainly used for production.

Since ancient times, red sea bream (*Pagrus major*) has been prized as the "king of fish" in Japan, because of its elegant appearance and color as well as its superior taste. It has become essential for celebrity meals for wedding ceremonies as an "auspicious fish." Its production value accounts for 10% of the total value of aquaculture. The seed production technique was developed in the 1960s and the main production method is cage culture. It is commonly cultivated in the western parts of Japan, especially around Kyushu Island and in the Seto Inland Sea.

An artificial hatching technique was developed in 1965 for the bastard halibut (*Paralichthys olivaceus*), and this was followed by commercialization in 1977. Production has dramatically risen since 1985, from 648 ton in 1983 to 3,097 ton in 1988 and then 6,039 ton in 1990. Inland aquaculture of the bastard halibut has also been making progress in recent years. In addition to the above species, Japanese horse mackerel (*Trachurus japonicus*), striped jack (*Pseudocaranx dentex*), and tiger puffers (*Fugu rubripe*) can be listed as the other main marine fish species cultivated.

For shellfish aquaculture, the yezo scallop (*Patinopecten yessoensis*) is one of the main species in coastal aquaculture in the northern regions of Japan. It has been exported to China for more than 200 years. Although in the past it was mainly bred using seabed-sown cultivation, success in hanging aquaculture in 1958 and subsequent technological advances have greatly increased its production.

Seabed sowing cultivation of oyster (*Crassostrea gigas, C. nippona*) has been performed in the tidelands of the Seto Inland Sea since the middle of the sixteenth century. However, the practical use of hanging culture with rafts was promoted in the 1920s, resulting in the expansion of production areas to areas other than tidelands. The production of oysters dramatically increased after aquaculture became possible in offshore fisheries by the adoption of hanging culture with longlines in the first half of the 1950s. Currently, *Crassostrea gigas* and *C. nippona* are the main species in shellfish aquaculture.

Among other shellfish cultured in Japan, there are abalone (*Haliotis discus*), Japanese carpet shell (*Ruditapes philippinarum*), Japanese scallop (*Pecten albicans*), and tunicate species (Tunicata).

Seaweed cultivation of nori seaweed (*Porphyra spp.*) was started by fishers in Tokyo Bay at the end of the seventeenth century using a support system. The success of artificial seed production in 1952 led to the expansion of production areas across the country. It is one of the main species in Japan's aquaculture industry, accounting for 28% of the volume of production and 20% of the value produced by aquaculture in Japan.

For sea mustard, or wakame (*Undaria pinnatifida*), technological development was encouraged by research institutes and fishery industries in the 1950s, leading to the commercial cultivation of sea mustard from around 1965. It currently accounts for 5% of total aquaculture production and 2% of the total value produced in Japan. Kelp, or konbu (*Laminaria japonica, L. angustata, L. longissima, L. ochotensis*), and *Cladosiphon okamuranus*, are also cultured.

For kuruma prawn (*Marsupenaeus japonicus*), hatching and raising in ponds started in 1889. Artificial seeds were introduced in 1963, which led to increased production. These aquaculture techniques have been transferred to China, Southeast Asia, India, and Latin America.

Another important cultured species is pearls. Kokichi Mikimoto was the first to succeed in pearl culture in 1893, and the technical developments needed for producing full-orbed pearls soon followed. After 1910, the production of full-orbed pearls was enabled by technical advances which have since been adopted by pearl farmers worldwide. These days, pearls are raised by hanging aquaculture. The production value accounts for approximately 5% of the total value of aquaculture, and they are mostly exported.

3.3.3.2 Freshwater Species (Takashima and Murai 2005)

Eel (*Anguilla spp.*) is the main species in freshwater aquaculture, and accounts for 40% of total production both in amount and value. Almost 100% of domestic eel is produced by aquaculture. The production of *Anguilla japonica* was first commercialized in 1879 and then was developed mainly in the central regions of the Pacific Coast. It is now also actively conducted in the western region of Japan. In recent years, it is thought that *Anguilla anguilla* and *A. rostrata elvers* from Europe and the United States, respectively, have escaped into the wild in Japan. As for fish seed, wild elvers are commonly captured for use, but in 2003, the National Research Institute of Aquaculture (NRIA) succeeded with seed production.

Commercialization of ayu (sweetfish: *Plecoglossus altivelis*) has progressed since the 1960s. Currently, ayu are released into rivers for use by commercial and recreational fisheries. Ayu production accounts for approximately 20% of the total amount earned in freshwater aquaculture.

For rainbow trout (*Oncorhynchus mykiss*), 10,000 eggs were brought in from California in 1977. Currently, the culture of rainbow trout is commercialized across the country and they have been released into numerous rivers. The culture of common carp (*Cyprinus carpio*) began in the Edo era. Ornamental carp (or koi), which are species with various color mutations, are also produced for display.

3.3.4 *Systems of Aquaculture*

3.3.4.1 Aquaculture Without Feeding

Seabed cultivation has long been employed as a method for clam culture. Spat collected from natural waters or produced in an artificial manner are disseminated in shallow sea areas. It is effective to prepare the growth environment prior to their dissemination, preventing the intrusion of, or removing, any predator. Given a good location, seabed cultivation can be very highly productive, as in the case of giant ezo scallop culture along the Okhotsk coast in Hokkaido.

Stakes or stick-method aquaculture is using bamboo or synthetic resin poles driven into the bottom of shallow sea areas. This method is primarily used for oyster and nori (laver) culture, since it allows a three-dimensional utilization of space up to the surface. Meanwhile, the "net method" is used for nori and other seaweeds that require sufficient sunlight for photosynthesis.

In hanging aquaculture, baskets containing shellfish, or shells, ropes, or other materials to which cultured shellfish or seaweed species are attached are lowered into the seawater from rafts or long lines fixed to the bottom with anchors. This method has the merit of allowing the utilization of greater areas of water for aquaculture than with other methods.

3.3.4.2 Aquaculture by Feeding

The most traditional method of aquaculture by feeding involves the use of natural or irrigation ponds for rearing fry. Aquaculture in farm ponds is used to raise freshwater fish, including carp. In addition to the need to feed the fish directly, aquaculture in farm ponds may require fertilization to propagate zooplankton or phytoplankton in the ponds to serve as feed.

Paddy culture makes use of paddy fields during the flooded period. Since the Meiji era, this method has been utilized for carp culture in various parts of the country. However, it is rarely seen today.

The commonest system is now cage aquaculture. It uses corves, which are facilities mainly comprising cubic or cylindrical net cages, frameworks to maintain their shape (rafts), plus floats and anchors to keep them on the water's surface. The resulting high exchange rate of water, which allows intensive fish culture, is a major contributor to the high productivity seen with this method. More details of aquaculture systems in Japan can be found in Honma (1993).

3.4 Fish Ranching and Processing

3.4.1 Fish Ranching

Fish ranching is the artificial production of fish seeds which are then released into natural waters, followed by appropriate management of their habitats and catches. It aims to ensure sustainable and efficient exploitation of fish species by protecting them against depletion of their spawn and larvae. For these species whose stocks are deteriorating due to overfishing, fish ranching is used as an emergency measure for stock recovery.

Fish ranching can be basically divided into two stages: seed development and site management. Seed development is the process of artificial rearing of spawn into larvae (seed production) and their release into the wild. On the other hand, site management entails the creation of suitable environmental conditions for the growth of released fish and to manage fishing operations at the site. Fish ranching has focused on species with low stock levels or high market value, and is often adopted alongside Resource Recovery Plans (Sect. 2.4.2).

The basic technologies for seed production have been developed by the National Centers for Stock Enhancement, which have been established across Japan since

Table 3.5 Major fish seeds released in fiscal year 2006

Species	Number of artificial fish seeds released (in thousands)
Red sea bream	17,518
Flounder	3,920
Bastard halibut	25,201
Kuruma prawn	122,422
Offshore greasyback prawn	35,053
Horse crab	32,095
Abalone	24,718
Ark shell	1,848
Yezo giant scallop	3,138,770
Spiny top shell	2,853
Sea urchin	59,604

Source: Fisheries Yearbook Editorial Committee (2009)

1963 (and were integrated into the Fisheries Research Agency in 2003). Their main tasks include developing technologies for stable seed production, investigating the effects on ecosystems caused by fish ranching, and developing technology for seed production of new species.

High-volume seed production has been conducted by the local ranching stations established in each prefecture since 1973. As of 2009, there were 69 centers across Japan. The fish seeds are sold to FCAs or local governments, and then released into the natural environment. Table 3.5 shows the number of major species released in FY 2006. Site management is the responsibility of local FCAs in collaboration with local governments and local ranching stations.

The key issue in fish ranching in Japan is the scientific assessment of both the economic and resource effects. A coordinated system that links fish ranchers in neighboring prefectures also needs to be established to ensure the effects of fish seed release on migratory species.

3.4.2 Processing

The fisheries processing sector is also an important sector in the Japanese fisheries industry. Because catch amounts fluctuate markedly according to the seasons and conditions in fishing grounds, the processing sector plays a critical role in efficiently utilizing landed fish and providing a steady supply of fisheries products to the Japanese population. The processing sector is, in face, the largest recipient of products from production sites (Table 3.6). Processing allows fisheries products to ensure good keeping qualities and find broader markets.

The processing sector also creates numerous jobs and adds value. In 2008, there were 178,000 employees in the fisheries processing sector, with productivity worth JPY 3,152 billion (Ministry of Economy, Trade and Industry 2009). The social and

Table 3.6 Balance sheet of fisheries products in 2007 (excluding seaweed and whales)

Items	Amounts (in 1,000 ton)
<Supply>	
Domestic production	508
Imports	516
<Demand>	
Fresh and frozen	299
Processing (edible)	426
Processing (animal feeds and fertilizer)	171
Processing (fish feeds)	57
Exports	81
Inventory change	−10

Source: MAFF (2008)

economic importance of the fisheries processing sector is about the same as the fisheries production sector, including aquaculture. Most fisheries processing companies are small-scale businesses based in provincial cities, and use locally landed products. These local processors play an important role in conserving and creating a diverse local food culture in Japan.

There is a variety of types of processed fisheries products for human consumption. One of the most popular is Surimi (minced fish) products. There are many types of Surimi products all around Japan, based on the locally landed fish species and the traditional food culture in the area. Dried or salted-and-dried fish are also consumed on a weekly basis. Mackerel, squid, sardines, anchovy, herrings, flounder, cod, tuna, scallops, kelp, etc., are popularly used in this processing technique. Canned products such as tuna or crab are also highly developed and are exported to other countries.

References

Akimichi T (1990) Small-type coastal whaling in Japan: report of an International Workshop. University of Alberta Press, Alberta

Baumgartner TR, Soutar A, Ferreira-Bartrina V (1992) Reconstruction of the history of the Pacific sardine and northern anchovy populations over the past two millenia from sediments of the Santa Barbara basin. Calif CALCOFI Report 33:24–40

Delgado CL, Wada N, Rosegrant MW, Meijer S, Ahmed M (2003) Fish to 2020: supply and demand in changing global markets (World Fish Center Technical Report 62). International Food Polcy Research Institute and World Fish Center. http://www.ifpri.org/sites/default/files/publications/oc44.pdf. Accessed 30 June 2010

Endo A, Yamao M (2007) Policied governing the distribution of by-products from scientific and small-scale coastal whaling in Japan. Mar Policy 31:169–181

Environment Agency and Marine Parks Center of Japan (1994) The report of the marine biotic environment survey in the 4th national survey on the natural environment. http://www.biodic.go.jp/reports/4-12/r00a.html. Accessed 30 June 2010 (in Japanese)

Fisheries Agency (2009) Fishing vessel statistics. Nourin-toukei-kyokai, Tokyo (in Japanese)

Fisheries Agency and Fisheries Research Agency (2009) Assessment of the fishery resources at the waters surrounding Japan. http://abchan.job.affrc.go.jp/digests21/index.html. Accessed 30 June 2010 (in Japanese)

Fisheries Yearbook editorial committee (2009) Fisheries yearbook of 2009. Suisansya, Tokyo (in Japanese)

Food and Agricultural Organization of the United Nations (2005–2010) National aquaculture legislation overview, Japan. (Text by Spreij, M.). http://www.fao.org/fishery/legalframework/nalo_japan/en. Accessed 30 June 2010

Food and Agricultural Organization of the United Nations (2008a) The state of world fisheries and aquaculture. Food and Agricultural Organization of the United Nations, Rome

Food and Agricultural Organization of the United Nations (2008b) Report of the FAO workshop on vulnerable ecosystems and destructive fishing in deep-sea fisheries (FAO Fisheries Report No. 829). Food and Agricultural Organization of the United Nations, Rome

Honma A (1993) Aquaculture in Japan. Japan-FAO Association, Tokyo

Kawasaki T (1983) Why do some pelagic fishes have wide fluctuations in their numbers? In: Csirke J, Sharp GD (eds.) Report of the expert consultation to examine changes in abundance and species composition of neritic fish resources (FAO Fisheries Report 291). Food and Agricultural Organization of the United Nations, Rome, pp 1065–1080

Kumai H (2005) Fisheries enhancement and aquaculture system (vol 1 Marine Fish). Kouseisya-Kouseikaku, Tokyo (in Japanese)

Makino M, Matsuda H (2011) Ecosystem-based management in the Asia-Pacific. In: Ommar R, Perry I, Curry P, Cockrane K (eds.) Coping with global changes in social-ecological systems. Wiley-Blackwells, London, pp 322–333

Matsuda H, Makino M, Tomiyama M, Glecich S, Castilla JC (2010) Fisheries management in Japan. Ecol Res 25(5):899–907

Ministry of Agriculture, Forestry and Fisheries (1960–2009) Annual statistics of fisheries and aquaculture production. Nourin-toukei-kyokai, Tokyo (in Japanese)

Ministry of Agriculture, Forestry and Fisheries (2008) Food balance sheet of 2007. Nourin-toukei-kyokai, Tokyo (in Japanese)

Ministry of Agriculture, Forestry and Fisheries (2009a) Annual statistics of fisheries and aquaculture production. Nourin-toukei-kyokai, Tokyo (in Japanese)

Ministry of Agriculture, Forestry and Fisheries (2009b) Annual report on fisheries labor survey of 2008. Nourin-toukei-kyokai, Tokyo (in Japanese)

Ministry of Agriculture, Forestry and Fisheries (2009c) Annual report on fisheries business survey of 2008. Nourin-toukei-kyokai, Tokyo (in Japanese)

Ministry of Agriculture, Forestry and Fisheries (2010) Fisheries census of 2008. Ministry of Agriculture, Forestry and Fisheries, Tokyo (in Japanese)

Ministry of Economy, Trade and Industry (2009) Statistical table of Japanese industry. Keizai-sangyou-chosa-kai, Tokyo (in Japanese)

Ministry of Finance (2009) Trade statistics. Nihon-kanzei-kyokai, Tokyo (in Japanese)

Mori K (2005) Fisheries enhancement and aquaculture system (vol 3 Shellfish, Crustacea, Echinus, and Seaweed). Kouseisya-Kouseikaku, Tokyo (in Japanese)

Morikawa J (2009) Whaling in Japan: power, politics and diplomacy. C Hurst & Co Publishers, London

Myers RA, Worm B (2003) Rapid worldwide depletion of predatory fish communities. Nature 423:280–283

Norin-Chukin Research Institute (2008) Present situation and problems of fisheries as indicated by questionnaire survey of fishers. http://nochuri.co.jp/report/pdf/n0807jo1.pdf. Accessed 30 June 2010 (in Japanese)

Ohshima Y (1994) History of fisheries enhancement and aquaculture technology development. Midorishobo, Tokyo (in Japanese)

Pauly D, Watson R (2005) Background and interpretation of the 'Marine Trophic Index' as a measure of biodiversity. Philos Trans R Soc Biol Sci 360:415–423

Pauly D, Christensen V, Dalsgaard J, Froese R, Torres F (1998) Fishing down marine food webs. Science 279:860–863

Takashima F, Murai M (2005) Fisheries enhancement and aquaculture system (vol 2 Freshwater fish). Kouseisya-Kouseikaku, Tokyo (in Japanese)

Watanabe Y, Zenitani H, Kimura R (1995) Population decline of the Japanese sardine Sardinops melanostictus owing to recruitment failures. Can J Fish Aquat Sci 52:1609–1616

Worm B, Barbier EB, Beaumont N, Duffy JE, Folke C, Halpern BS, Jackson JBC, Lotze HK, Micheli F, Palumbi SR, Sala E, Selkoe KA, Stachowicz JJ, Watson R (2006) Impacts of biodiversity loss on Ocean ecosystem services. Science 314:787–790

Yatsu A, Watanabe T, Ishida M, Sugisaki H, Jacobson LD (2005) Environmental effects on recruitment and productivity of Japanese sardine *Sardinops melanostictus* and chub mackerel *Scomber japonicas* with recommendations for management. Fish Oceanogr 14:263–278

Chapter 4
Fisheries Management in Coastal Areas

Abstract Three cases from coastal fisheries management are described here. At Mutsu Bay, located in the northern part of the main island, high-priced sea cucumbers are managed by an organization established within a local Fisheries Cooperative Association (FCA). It implements wide-ranging measures that include no-take zones and TACs based on the results of resource assessments conducted by the fishers themselves. The second case is the sand eel fishery in Ise Bay, in the central part of the main island. Fishers from 12 FCAs in two prefectures surrounding the bay share the same stock for different uses. After a long period of conflict between fishers and the grim lesson of resource collapse, they implemented a highly science-based management strategy that continues to deliver successful results. Third, sandfish migrate throughout the Japan Sea, and are shared by coastal and offshore fishers from four Prefectures. After huge efforts and more than 200 meetings among fishers, local government officers, and researchers, fishers in Akita Prefecture finally adopted a 3-year moratorium that gave successful results. The management regime was expanded to include all four Prefectures, and sandfish stock is now officially managed under a Resource Recovery Plan.

4.1 Introduction

Japanese coastal fisheries have a highly complex setup, and their resource uses are quite intensive. Table 4.1 shows an example of the variety of fishing gear used and the target species at Kurahashi-jima community in the 1930s. Within one community, 21 types of equipment and 44 types of fishing practices were officially defined and managed by the cooperative's rules. These were allocated to the 739 fishers living in the community.

Today, more than 186,000 fishers, or about 87% of the total in Japan, are coastal fishers. Due to the complexity of the system and its intensive nature, fisheries coordination and resource conservation cannot be implemented effectively in a

M. Makino, *Fisheries Management in Japan: Its institutional features and case studies*, Fish & Fisheries Series 34, DOI 10.1007/978-94-007-1777-0_4,

Table 4.1 Types of coastal fisheries at Kurahashi-jima community in the 1930s

Gear type		Target	Number of fishers
Net fishery	Boat seine	Sardine	49
		Miscellaneous	16
		Waina	3
	Beach seine	Mullet	3
		Miscellaneous	13
		Halfbeak	1
	Kieta-net		5
	Dredge-net	Keta-ami	5
		Shellfish	27
	Mashu-net		20
	Dip net	*Yahagi*	7
		Miscellaneous	1
	Sailing trawl		10
	Show-ami	Miscellaneous	19
		Filefish	29
	Gochi net		5
	Trap	*Kochi*	9
		Sea bream	1
		Chinu	2
		Wachi	5
		Amate	2
		Flatfish	3
		Gori	2
	Danish seine	Miscellaneous	30
		Squid	1
	Kogi-net	Sea cucumber	38
		Shrimp	1
	Drift net	Jack mackerel	1
		Spanish jack mackerel	36
	Tsubo-net		19
	Shibari-net		12
	Tempo-net		1
	Katsura-net	Sea bream	2
	Set-net fishery	Black rockfish	1
Longline fishery		Sea bream	34
		Miscellaneous	15
		Pike eel	12
		Yellowtail	7
		Flatfish	12
Pole and line	Chum fishery		145
		Miscellaneous	28
		Black rockfish	13
		Sea bream	14
Pot		Octopus	81
Aquaculture		Oyster	4
Total			739

Modified from Yamaguchi (2007)

top-down, command-and-control manner. As all the three cases presented in this chapter clearly show, fisheries management organizations (FMOs, Table 2.2) consisting of allied fishers, play a core role in decision making, and then implementing and enforcing the measures they decide on their own initiative. Local and central government officials and researchers support FMOs' activities with legal advice and scientific information.

My first case is a sea cucumber fishery in Mutsu Bay. This highly priced sedentary species is managed by an FMO organized within the local FCA. It autonomously adopts various resource conservation measures such as minimum and maximum size limits, no-take zones, and annual allowable catches based on resource assessments conducted by the fishers themselves. This FMO also directed a lot of effort toward perfecting dry processing techniques and marketing the sea cucumbers for export to Hong Kong. Their dried sea cucumber product is now one of the most recognized seafood commodity brands in the Hong Kong markets.

The second case is a sand eel fishery in Ise Bay. Sand eels estivate and spawn at the mouth of the bay, then migrate into the bay, where they are harvested. Fishers from 12 FCAs in two prefectures surrounding the bay share the same stock. Historically, fishers from one prefecture have mainly targeted juveniles, while fishers from the other have harvested adult sand eels. After a long period of discord, followed by a severe stock collapse in the early 1980s, fishers organized a cross-prefectural FMO and autonomously adopted highly science-based measures for sustainable resource use.

The third case concerns sandfish resources in the northern Japan Sea. This species migrates across four prefectures, and is harvested by both coastal and offshore fisheries. An FMO organized by fishers in Akita took the initiative in calling a moratorium on sandfish for 3 years, which achieved a successful result. Then, with legal support from the central government, a large FMO consisting of both coastal and offshore fishers across four Prefectures was finally organized in 2003.

Chapter 5 has two case studies of offshore fisheries. A discussion on all five fisheries regarding both coastal and offshore fisheries management is included in the concluding chapter (Chap. 10).

4.2 Sea Cucumber Fishery in Mutsu Bay

4.2.1 Background

Since around the seventeenth century, dried sea cucumber (*Aposticopus japonicus*) from the northern part of Japan has been highly valued in the Chinese market. According to a record from 1745, the annual export weights of dried sea cucumber and dried abalone were 190.46 and 79.84 ton, respectively. Dried sea cucumber was a major Japanese export commodity, along with dried abalone, dried shark fin, dried fish bladder, etc. They were called "Tawara-mono," meaning dried commodities enveloped in straw bags (tawara) for export. At the time, Tawara-mono were an important source of foreign currency for Japan. From the late eighteenth century,

Table 4.2 Dried sea cucumber exports from Japan

Year	Volume (1,000 ton)	Value (JPY billion)
2004	223	5.5
2005	230	7.8
2006	273	12.6
2007	345	16.7
2008	283	13.3

Source: Ministry of Finance (2005–2009)

the feudal government directly managed the whole production and transportation process of dried sea cucumbers.

The first record of dried sea cucumber production in Mutsu Bay, which is located at the northern tip of the main island of Japan (Fig. 1.1), was in 1653. Since then, Mutsu Bay has been one of the most famous sea cucumber production sites in Japan. In the late eighteenth century, sea cucumber fisheries were officially managed by the feudal lord, and the representatives of local sea cucumber fishers were allowed to hold swords and family names, which means they were regarded as a kind of Samurai, a noble class.

Until recently, dried sea cucumbers were produced individually by many local fishers in Mutsu Bay. However, harvesting and drying sea cucumbers is a long and laborious process, and for the last several decades, scallop aquaculture, which is much less arduous and more profitable, has been more common throughout Mutsu Bay. During this period, small numbers of sea cucumbers were landed, but only for domestic consumption.

Around 1995, the price of cultured scallops began to decline, due to growing scallop production volumes in other areas of Japan. Also, probably due to overcultivation in the bay, annual scallop production volumes fluctuate drastically. On the other hand, the price of dried sea cucumber in the Hong Kong market has increased sharply in parallel with economic growth in China. Many Mutsu Bay fishers therefore resumed sea cucumber fishing at the end of the 1990s (Hirota 2011). Table 4.2 shows the official statistics for dried sea cucumber exports from Japan, which started in 2004.

The Kawauchi-machi Fisheries Cooperative Association (FCA) is located on the eastern side of Mutsu Bay. At the end of the 1990s, this FCA adopted wide-ranging measures governing sea cucumber fishing, and its products are now famous in the Hong Kong market for their high standard of resource conservation and their dry processing quality. In 2004, the Emperor's Prize was awarded to the Kawauchi-machi FCA for their high-standard autonomous activities in fisheries management (Makino 2011).

4.2.2 Sea Cucumber Fisheries

Fishers in Kawauchi-machi FCA use small dredge nets for catching sea cucumber (Fig. 4.1). This is categorized as a governor-licensed fishery in Aomori Prefecture.

Fig. 4.1 Dredge net hauled onto the back deck

In 2007, there were 144 fishers in the Kawauchi-machi FCA, most of whom were engaged in sea cucumber dredge-net fishing. The FCA itself also operates a diving fishery based on the fishery rights vested to it. These two types of gear are the main ones used for sea cucumber fishing in this FCA. Table 4.3 summarizes the changes in annual production by the Kawauchi-machi FCA. Values are for raw sea cucumber. As can be readily seen, the production volume has increased since the end of the 1990s, and the unit price has sharply increased since 2003.

About half of landed sea cucumbers are sold fresh on the domestic market, and the other half are processed into dried sea cucumbers for export to Hong Kong. The drying process is conducted by an FCA-direct factory under strict quality control. Their products have a good reputation, and are now a recognized brand in the Hong Kong seafood market (Fig. 4.2).

4.2.3 Management

4.2.3.1 FMO Structure

In 1999, the Kawauchi-machi FCA organized an FMO within the FCA, named the Council for Promoting Sea Cucumber Resource Utilization (Fig. 4.3). This council has played a core role in autonomous management activities by the member fishers. The council's authority covers resource assessment of sea cucumber stock in their

Table 4.3 Productions of sea cucumber by the Kawauchi-machi FCA

Year	Volume (ton)	Value (JPY thousands)	Unit price (JPY/kg)
1981	115	47,348	412
1982	128	63,769	498
1983	83	43,702	527
1984	93	36,743	395
1985	72	40,627	564
1986	16	7,097	444
1987	28	13,668	488
1988	38	21,015	553
1989	90	41,451	461
1990	31	12,289	396
1991	188	62,912	335
1992	236	97,092	411
1993	79	36,211	458
1994	64	25,076	392
1995	94	29,770	317
1996	61	18,935	310
1997	223	77,503	348
1998	49	15,331	313
1999	276	125,854	456
2000	235	132,791	565
2001	276	129,948	471
2002	311	150,455	484
2003	335	253,157	756
2004	421	409,732	973
2005	269	315,697	1,174
2006	299	572,365	1,914
2007	239	512,244	2,143
2008	253	418,971	1,656

Source: Kawauchi-machi FCA (1982–2009)

fishing grounds, planning of the annual harvest based on the results of resource assessment, dry processing at the FCA-direct factory within the FCA site, and promotion activities to the domestic and Hong Kong markets.

4.2.3.2 Management Measures

Every dredge net vessel operating in Mutsu Bay must have a license issued by the prefectural governor. The Prefectural Fishery Coordinating Regulations (Table 2.2) state that dredge fishing in Mutsu Bay may come no nearer than 500 m from the coast. The Regulations also limit their fishing season to October to April with operation times from 7 am to 11 am. In addition to these official management measures, a wide range of measures are autonomously implemented by the strong leadership of the Council for Promoting Sea Cucumber Resource Utilization.

Fig. 4.2 Dried sea cucumbers produced by *Kawauchi-machi* FCA

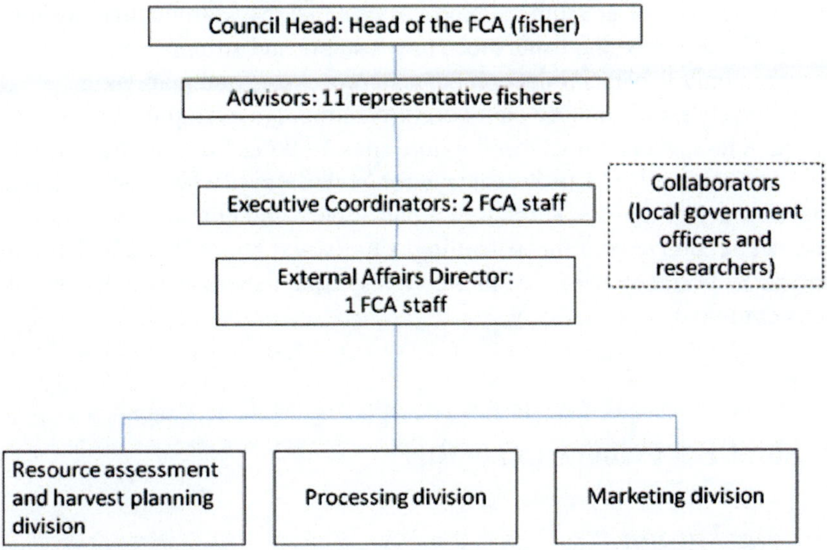

Fig. 4.3 Structure of the Council for Promoting Sea Cucumber Resource Utilization within the *Kawauchi-machi* FCA

As mentioned earlier, fishers have conducted annual resource assessments since 1998 with scientific support from the prefecture fisheries research station. They set the Total Aallowable Catch (TAC) of the year based on the assessment results. More recently, the annual TACs have been set at around 60% of total biomass, taking into account the conditions of Hong Kong markets and other operations such as scallop aquaculture.

They also set the catch size limit and daily catch limit based on directions from the Council. The minimum size limit is presently 120 g and the maximum size limit is 300 g, taking into account both the reproductive maturity of the sea cucumber and Chinese consumer demand. To optimize the working rate of the FCA-direct processing factory, the daily catch limit is set to 50–60 kg per vessel per day.

As an input control strategy to reduce overall fishing pressure, the fishers form four groups of dredge net vessels, with each group operating in turn. Artificial reefs made of scallop shells are built to enhance the sea cucumber habitat, especially to protect juveniles and spawners. The appropriate shapes, locations, and depths of the reefs are advised by the prefecture fisheries research station, and the areas around the reefs are designated by the Council as no-take zones. In addition, the Council is now taking the initiative in conducting research and experimentation on producing artificial sea cucumber seeds for ranching in their fishing grounds.

In this case, an effective FMO (a Council for Promoting Sea Cucumber Resource Utilization) could be organized within one FCA, since the target species is sedentary. Also, because this site has been a famous sea cucumber production site for hundreds of years, traditional ecological knowledge of local sea cucumber resources and mutual trust among fishers have been built up over generations. These conditions smoothed the launching of the new FMO. Also, specific economic conditions—a sharp increase in the sea cucumber export price and downward fluctuations in scallop production—motivated the fishers to set up a stable and strong management regime covering not only fishery production but also processing and marketing.

There are still several issues left here. Poaching, for example, is a serious problem at all the sea cucumber production sites in Japan due to the recent sharp increase in export price. This is also true for Mutsu Bay. To deal with this issue, the Kawauchi-machi FCA has purchased a patrol boat and employs two staff to deter poachers. The total cost of this patrolling activity was about JPY 5 million in 2008, which was all borne by the FCA. In the high season, fishers also patrol their fishing grounds at night.

4.3 Sand Eel Fishery in Ise Bay

4.3.1 Background

Ise Bay is located on the Pacific side of the central part of Japan, bordered by Aichi and Mie Prefectures (Fig. 1.1). Two-boat pelagic trawl fishers from both Prefectures harvest Japanese sand eels (*Ammodytes personatus*). Based on the local coordination and negotiation history between the two Prefectures, fishers from Aichi Prefecture now mainly target juveniles, while Mie fishers mainly target adult sand eels (Table 4.4).

Table 4.4 Changes in sand eel production volume and values in Ise Bay

	Volume (ton)			Value (JPY thousand)		
Year	Aichi Pre.	Mie Pre.	Total	Aichi Pre.	Mie Pre.	Total
1979	1,619	352	1,971	7,048	352	7,401
1980	1,352	2,031	3,383	3,435	138	3,573
1981	848	606	1,454	2,368	1,614	3,982
1982	343	172	515	1,455	801	2,257
1983	3,889	5,323	9,212	5,815	6,683	12,498
1984	3,774	1,501	5,275	5,623	2,232	7,856
1985	4,619	6,988	11,607	8,565	6,186	14,751
1986	5,950	6,346	12,296	12,165	7,710	19,875
1987	4,559	5,179	9,738	8,065	4,393	12,458
1988	4,195	2,719	6,914	9,262	5,213	14,475
1989	4,553	3,181	7,734	10,931	4,239	15,170
1990	1,588	832	2,420	6,340	2,534	8,875
1991	2,582	2,647	5,229	12,194	5,626	17,820
1992	11,301	14,358	25,659	11,186	6,786	17,972
1993	7,559	8,077	15,636	10,531	6,592	17,123
1994	2,970	4,471	7,441	15,504	9,219	24,724
1995	1,875	1,160	3,035	5,581	2,322	7,904
1996	5,883	5,022	10,905	18,154	9,209	27,364
1997	4,081	4,052	8,133	8,253	5,388	13,641
1998	797	397	1,194	4,901	1,897	6,798
1999	4,450	5,995	10,445	8,202	6,522	14,724
2000	559	356	915	3,854	1,918	5,771
2001	5,688	8,965	14,653	9,241	7,374	16,616
2002	7,127	9,349	16,476	10,252	7,404	17,656
2003	3,120	1,715	4,835	9,362	3,724	13,086
2004	10,737	8,372	19,109	9,983	6,525	16,508
2005	3,972	4,980	8,952	6,144	3,825	9,969
2006	8,528	10,545	19,073	7,844	5,329	13,174

Source: Tomiyama et al. (2008)

Since around 1960, technological progress, such as greater engine power, bigger fishing gear, and the installation of echo sounders has brought dramatic increases in the fishing capacity of pelagic trawlers in the bay. At the same time, during this period, there was major growth in fish seed production technologies for aquaculture throughout Japan, and demand for adult sand eel as fish meal rapidly increased. These two factors led to intensive exploitation of sand eel resources in the late 1960s and the early 1970s.

The sand eel catch in Ise Bay then dramatically declined in the early 1980s, most likely due to overfishing. After the collapse of the stock, scientific management measures were introduced through collaboration between the local fishers and researchers. As a result, the total harvests of Aichi and Mie Prefectures improved from 515 metric tons (JPY 2.26 million) in 1982 to 19,073 ton (JPY 131.74 million)

Fig. 4.4 Pelagic trawl fleet for sand eel fishery (*fishing boats: left* and *center*, and *transporting boat: right*)

in 2006 (Tomiyama et al. 2008). This fishery was awarded as Marine Eco-Label Japan (MEL-Japan) certificate in 2010 (MEL-Japan 2010).

Sand eel larvae hatch at the mouth of Ise Bay in December and January. Only 3–4 mm in length, they are swept into the head of the bay by currents. Around May, when the bottom water temperature rises above 17–20°C, the sand eels return to the mouth of the bay and burrow into the bottom substrates to estivate. Their body length at this time is about 7–10 cm. The estivation grounds occupy wide areas at a depth of 20–50 m along the coast. The estivation of the sand eel ends and maturation starts around the end of November, when the water temperature falls below 15°C. In mid-December, they start to spawn at the mouth of Ise Bay. The life span of a sand eel is 3 years, and maximum body length is almost 16 cm (Tomiyama and Yanagibashi 2004).

4.3.2 Sand Eel Fishery

Sand eel fishing in Ise Bay is conducted by pelagic-trawl fleets consisting of two-net boats (about 15 gross tons) equipped with net winches and one or two transporting boats (about 15 gross tons) (Fig. 4.4). Catches are transferred from the cod end of the net into plastic baskets holding crushed ice on the deck of the transport boat. While the transport boat brings the harvest to landing port, the two-net boats continue to harvest.

Fig. 4.5 (**a**) Juveniles and (**b**) adults of sand eel

The price of the catch is determined mainly by the size and color of the sand eels. There are three categories, defined by their body length: (1) juveniles with a body size of 3–5 cm, called "shirasu," caught in March and April; (2) juveniles over 6 cm long caught in April–May, and (3) adult fish caught in January and February of the following year (Fig. 4.5). The fishers in Aichi Prefecture mainly catch (1) for human consumption, while the majority of (2) and (3) are harvested by Mie Prefecture fishers, chiefly for use as fish meal for aquaculture.

About 200 fleets comprising 700 fishing vessels from Aichi and Mie Prefectures are currently utilizing the sand eel stock in the bay. All the fishers belong to 1 of the 12 Fisheries Cooperative Associations (FCAs) located along the Ise Bay coastline, where they land and anchor their boats. All the landings must be sold by auction and managed by local FCAs. As explained below, strict fishing seasons are implemented by fishers. In the off season, some sand eel fishers harvest anchovy (*Engraulis japonicus*), while others engage in small-scale bottom trawling or seaweed aquaculture.

4.3.3 Management

4.3.3.1 FMO Structure

After the severe collapse of sand eel stocks in the early 1980s, fishers in 12 FCAs from both prefectures organized a cross-prefectural organization as an FMO for sharing the sand eel stock in Ise Bay. Figure 4.6 shows the structure of this cross-prefectural FMO. First, based on the type of operation, fishers organized two fishers' unions (FUs) within each prefecture: a shirasu FU comprising fishers harvesting anchovy and sand eel juveniles; and a Pelagic Trawl FU for anchovy, sand eel, and

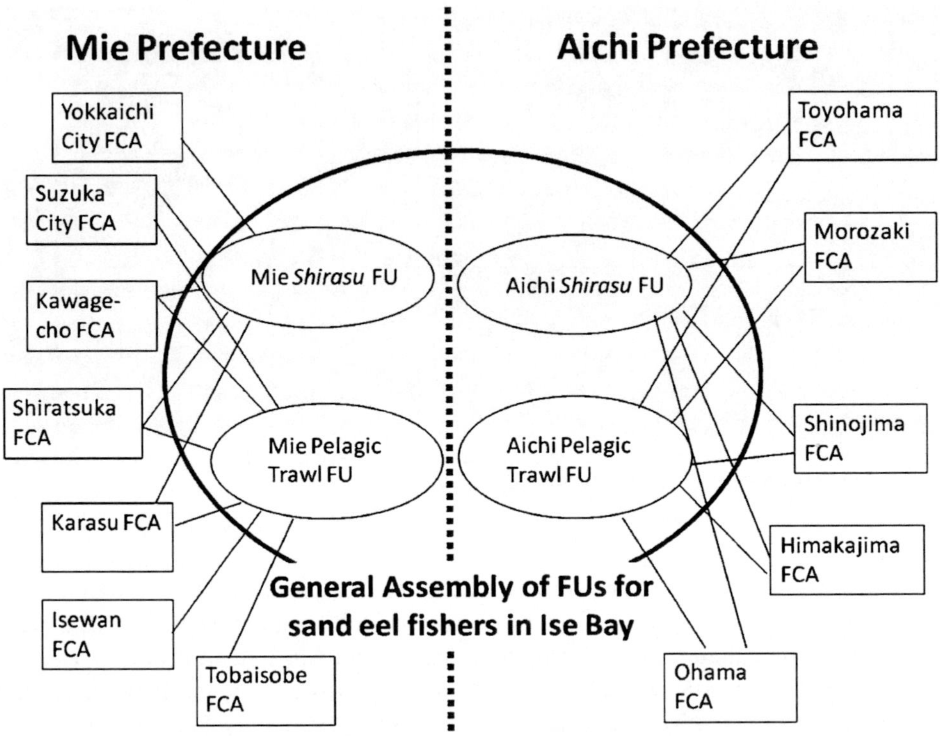

Fig. 4.6 Structure of the sand eel fishermen's organizations in Ise Bay

sardine fishers. These four unions are now organized as a general assembly of FUs. This framework plays the central role in decision making for the management of sand eel resources in Ise Bay.

4.3.3.2 Management Measures

Each pelagic trawl vessel must have a license for sand eel fishing issued by the governor of each prefecture. The Prefecture Fisheries Coordinating Regulations in each prefecture also regulate the fishing season, equipment to be used, areas, etc. In addition, a wide range of management measures have been implemented on an autonomous basis, as explained below.

After the stock collapse of the early 1980s, local fishers limited fishing operations before and after the sand eel estivation. Basically, these measures were designed to protect spawning stocks. On the other hand, researchers in Aichi and Mie Prefectures conducted research on sand eel fisheries in Ise Bay that focused mainly on stock management, mechanisms of sand eel stock fluctuation, and methods for effectively using the stock. Based on these studies, fishers and researchers cooperatively introduced science-based management measures in the early 1990s. There are three principal measures for sand eel resource conservation: establishment of an opening day for fishing juveniles, protection of spawning stock, and establishment of protected areas for estivation.

Setting the Opening Day for Fishing Juveniles

Every year, researchers estimate the relationship between body length and possible opening days of juvenile fishing in March and the expected catch value during the fishing season. Scientific surveys of sand eel larvae are also conducted by researchers approximately every 10 days from research vessels using bongo nets. Researchers then estimate the optimal opening date for harvesting juveniles, taking annual fluctuations in growth rate into account.

Based on this scientific information, fishers hold discussions within each prefecture to decide the opening day for juvenile fishing. Additional information on market demand is also considered. A General Assembly of Fishery Unions (Fig. 4.6) is then organized to make the final decision on the opening day. Local government officials and researchers from prefectural research stations also attend the General Assembly as observers. The venue of the General Assemblies is in Aichi and Mie Prefectures in turn, and very heated discussions are often held between the fishers in the two Prefectures. Usually, the Aichi fishers want to start harvesting juveniles earlier, while the Mie fishers call for a later start. According to a local government official who has observed in General Assemblies many times, fishers in both prefectures have been gradually building up mutual trust by repeated annual meetings, and as a result, the discussions in recent General Assemblies are much shorter than before.

Protection of Spawning Stock

Estivation is an ecological characteristic of sand eels. During this period, they stay under the sand without consuming any food. As the natural mortality during estivation is relatively stable, it is possible to manage the number of spawning adults by controlling the harvest before estivation. In other words, it is possible to manage the reproduction of the stock by choosing the closing date of the fishing season.

Researchers in both prefectures agree that a spawning stock level of two billion is necessary for sufficient reproduction and recruitment to ensure sustainable stock levels of sand eel overall. Based on the daily landing reports from 12 FCAs to the research stations, the researchers notify the representatives of the FUs that the fishing season is coming to an end, since the estimated number of reproductive-age sand eel stock is approaching two billion. The actual closing date is determined by discussions between the fishers in both Prefectures.

To keep a sufficient number of eggs for the next season, adult sand eel fishing should be resumed not earlier than the end of the spawning period. Therefore, every year, the end of spawning period is confirmed by scientific sampling, jointly conducted by researchers and fishers.

Protected Areas for Estivation

As they grow from larvae to juveniles to adults, sand eels migrate gradually from the interior toward the mouth of Ise Bay, where they begin estivation in May–June. To ensure protection of these aestivating stocks, no-fishing zones are established at the mouth of the bay based on the agreement made at the General Assembly. The total area of the no-fishing zone varies every year. If the estimated size of stock is smaller, the zone area is made larger. The locations and areas of the zone change adaptively in response to ecological information such as the distribution of juvenile sand eel and migration paths. No pelagic-trawl fishing is permitted within the zone until the start of sand eel estivation. The beginning of estivation is scientifically determined by researchers in May–June.

In this case, the most important factor in facilitating management was that fishers understood the need for resource conservation. The severe collapse in sand eel numbers and the economic losses suffered in the early 1980s provided a grim lesson for local fishers. To improve their understanding of the situation, Extension officer (Sect. 1.5.2) and researchers explained the scientific basis for management in easily understood terms. Data used for resource analysis are collaboratively collected by fishers and researchers, which similarly enhances the legitimacy of science-based management. This fishery is famous throughout Japan as having highly scientific fisheries management as well as for its close collaboration between fishers and researchers.

However, the mortality rate during the juvenile period is affected by environmental variation, such as the strength of ocean currents, so annual stock fluctuations are very wide. This inevitably leads to instability of total income for sand eel fishers. Therefore, since 2006, a Resource Recovery Plan (Sect. 2.4.2) has been implemented for the sand eel stock in Ise Bay. The objective of this plan is to stabilize the annual landings at an appropriate level. The nature of the Resource Recovery Plan in this case is the authorization of autonomous measures mentioned above.

4.4 Sandfish Fishery in Akita Prefecture

4.4.1 Background

Akita Prefecture is located in the northern part of the Japan Sea (Fig. 1.1). The name for sandfish (*Arctoscopus japonicus*) in Japanese literally means "god fish," and legend says this fish brings thunderstorms. Sandfish is the most famous winter fish in

Fig. 4.7 Mature female sandfish

Fig. 4.8 Sorting process of landed sandfish

Table 4.5 Changes in sandfish production volume and value in Akita Prefecture

Year	Volume (ton)	Value (JPY million)
1979	1,390	733
1980	1,919	912
1981	1,938	1,138
1982	1,244	853
1983	357	297
1984	74	71
1985	203	206
1986	373	475
1987	286	397
1988	248	430
1989	208	367
1990	150	322
1991	70	141
1992	40	39
1993	–	0
1994	–	0
1995	143	420
1996	244	476
1997	469	765
1998	589	612
1999	730	780
2000	1,085	745
2001	1,569	1,114
2002	2,112	933
2003	2,969	1,138
2004	3,258	1,029
2005	2,402	772
2006	2,625	804
2007	1,653	835

Source: MAFF (1982–2009)

this area and is designated as the Prefectural Fish. Local people say, "You cannot see in the New Year without a sandfish." It is deeply rooted in the local food culture: they have various types of sandfish dishes, boiled with vegetables or salted and grilled.

The catch of sandfish in Akita Prefecture exceeded 20,000 ton in the 1960s, but sharply decreased from around 1976, falling to 71 ton in 1991. During this period, even the local fishers had to buy sandfish from Hokkaido or Korea. Faced with this drastic decrease in the catch, fishers in Akita Prefecture autonomously implemented a 3-year moratorium on sandfish harvesting in 1992.

After a major effort, they were able to successfully restore the sandfish population. They also adopted various additional measures including Total Allowable Catches (TACs) and individual vessel quotas (IVQs). In recent years, the total annual catch has been around 2,000 ton (Table 4.5).

Sandfish is chiefly distributed in the Japan Sea, and there are two stocks identified by the official stock assessment framework: the northern Japan Sea Stock and the western Japan Sea Stock. Fishers in Akita Prefecture catch the northern Japan Sea stock. Sandfish has a life span of about 5 years, and the ages of sexual maturity are 1 year (6–12 cm) for males and 2 years (over 15 cm) for females. They usually live at a depth of about 200 m, and are harvested by bottom trawlers when 1 year old. In winter, mature females lay 600–2,500 eggs on sea grass growing in rocky areas about 2 m deep. Akita Prefecture is the main spawning ground for the northern Japan Sea stock. The mature females, which have a greater market value, are harvested by coastal fisheries.

Coastal fisheries harvest the majority of the catch, but some are caught offshore. In the late 1980s, each side suspected the other side of being responsible for the decline of their catch. This fish also migrates across four Prefectures in the northern part of the Japan Sea. These conditions made cross-jurisdictional and cross-sector coordination very difficult. However, scientific information provided by researchers and administrative support by the central government played an important role in building consensus among the fishers affected, and ultimately a large FMO was established that covered all the coastal and offshore fishers across four Prefectures.

4.4.2 Sandfish Fisheries

Akita Prefecture and other three Prefectures (Aomori, Yamagata, and Niigata Prefectures) share the northern Japan Sea Stock. In Akita Prefecture, three types of fisheries harvest sandfish. Most are harvested by two types of coastal fisheries: small-scale set net fishing and gillnet fishing. In the offshore area, bottom trawlers operate in areas about 200–300 m deep.

Mature sandfish migrate to coastal areas in winter, and coastal fishers harvest them in December and January. The season lasts only about a week, and during this period almost all coastal fishers are engaged in set net and gillnet fishing of sandfish.

4.4.3 Management

All the fisheries types targeting sandfish must have rights or licenses issued by the governor of Akita Prefecture. The number and size of vessels, fishing grounds, equipment size, mesh size, etc., are officially controlled by means of these rights and licenses. The Akita Prefectural Fisheries Coordinating Regulations also set a minimum-size limit, and prohibit the collection or selling of sandfish eggs in coastal areas.

As an autonomous management measure, a stricter minimum-size limit was implemented (14 cm) in November 1986 based on an agreement reached among the chairs of all the 12 FCAs harvesting sandfish in Akita Prefecture. Also, several additional measures, such as mesh-size enlargement or seasonal no-take zones were introduced. However, the resource status continued to worsen, so in February 1992, the chairs of the 12 FCAs in Akita Prefecture agreed that, to allow the sandfish population to recover, drastic measures, including a moratorium, must be implemented.

Many fishers at the time thought that the decline in sandfish catch was just a natural fluctuation and would reverse itself again in the future. However, local researchers provided a more pessimistic outlook based on simulation results showing the possible recovery paths under various management options, including a 3-year moratorium. Fishers in the 12 FCAs, researchers, local government officers, and Extension Officers then held more than 200 intensive discussions over the next 6 months. Finally, they reached the "Agreement for Sandfish Resource Management" on October 1, 1992. Based on this agreement, a 33-month moratorium on all sandfish fisheries operations in Akita Prefecture was implemented from October 1, 1992, to June 30, 1995. This was actually a hard decision, and how the consensus was built is analyzed in detail by Suenaga (2008).

One fortunate thing was that the sandfish fishers were able to find other sources of income during the moratorium. For example, from 1992, coastal fishers were provisionally allowed to engage in longline fishing, and surprisingly harvested large numbers of tiger puffer (*Takifugu rubripes*), and offshore fishers were lucky to catch a lot of blackmouth angler (*Lophiomus setigerus*). Both these species have a much higher market value than sandfish.

4.4.3.1 FMO for Post-moratorium Management

How to resume sandfish fishing operations after the moratorium was a big question. A leading local government officer at the time said that resuming operations in 1995 was a much harder job than adopting the moratorium in 1992. Fishing in the same way as during the pre-moratorium period would naturally undo the benefits of the 3-year moratorium. This meant that the post-moratorium operations should be managed under a new management framework with new regulations. Sandfish fishers, local government officers, and researchers again held intensive discussions, and implemented the following measures.

First, to reduce the overall fishing pressure on the population, the number of offshore trawlers was cut from 57 vessels to 38, and for the coastal fisheries there was a reduction of 20% in set nets and 40% in gill nets, with financial support from the government.

Second, a new FMO, named the Sandfish Resource Measures Council, composed of both coastal and offshore fishers from Akita Prefecture, was established to discuss and implement new management measures after the moratorium (Fig. 4.9). The Council coordinated the reduction in the number of boats, larger mesh sizes, setting minimum size limits, etc. It also implemented an autonomous Total Allowable Catch (TAC) for sandfish, at about half of the estimated biomass. Coastal fisheries are allocated 60% of the TAC, with 40% going to offshore fisheries. The allocated portion of the TAC is divided into annual catch quota and then reallocated to each FCA. Each FCA decides how to use their allocated catch quota. Several FCAs have adopted the individual vessel quota (IVQ) system, while others

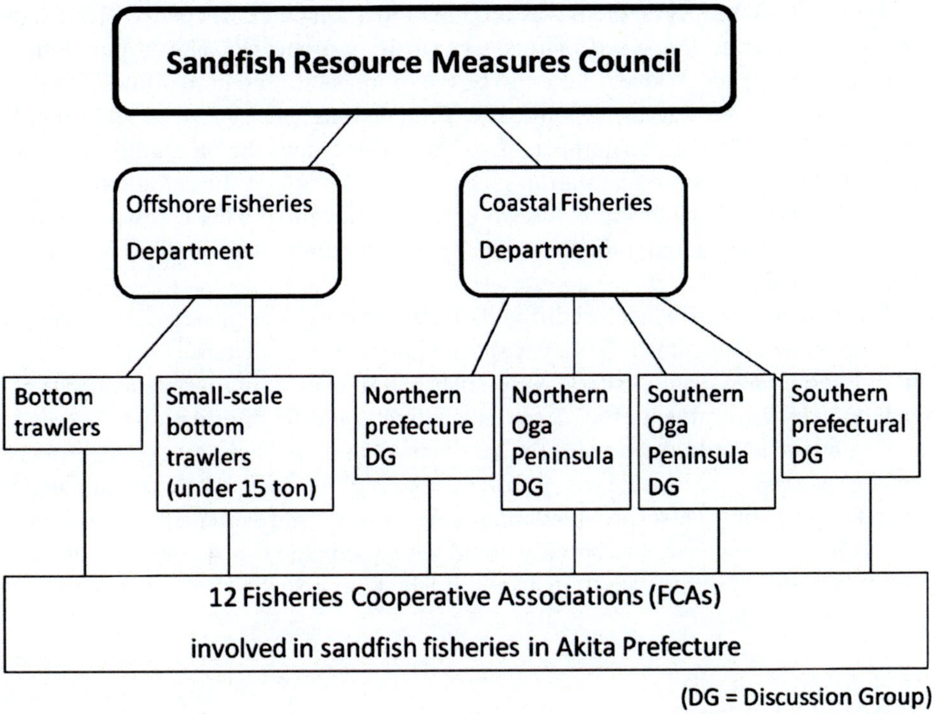

Fig. 4.9 Structure of the Sandfish Resource Measures Council in Akita Prefecture

have opted for a jointly operated regime. The rules on quota use are based on the FCA Regulations or FMO Rules (Tamaki 2004).

4.4.3.2 Enlarging the Geographical Scope

As pointed out earlier, the northern Japan Sea stock of sandfish is shared with fishers in four neighboring Prefectures: from north to south, Aomori, Akita, Yamagata, and Niigata. Therefore, to manage sandfish resources in a biologically consistent manner, it was crucial to establish a cross-prefectural management regime. However, the importance of sandfish was viewed differently by the fisheries in different prefectures, so their incentives for agreeing to additional regulations differed considerably. After a long process of coordination by local and central government, the fishers in the four prefectures finally signed up to the Sandfish Resource Management Agreement (see Sect. 2.4.3) in April 1999. Although there were differences in the content of the measures among Prefectures because of the specific operational conditions in each prefecture, the minimum size limit was standardized at 15 cm across the four prefectures. The management regime now covered all the fishers catching this species.

In July 2003, the Agreement was upgraded to a Resource Recovery Plan (Sect. 2.4.2) to accelerate the speed of recovery of the northern Japan Sea population. The plan was for the recovery target to be the biomass sufficient to allow 5,000 ton of this stock to be harvested annually. Total fishing pressure was additionally decreased by reducing the number of fishing vessels and the enlarging net mesh size. Another ecosystem restoration measure was to replant the sea grass beds in the coastal area of Akita Prefecture to enhance spawning. This cross-prefectural management scheme under the Resource Recovery Plan will be in effect for 9 years and last until 2011.

This case was one of the most difficult of the many fisheries management attempts undertaken in Japan, since the target species migrate long distances across coastal and offshore areas and fishers were distributed over four prefectures. Strong leadership by government officers and presenting scientific information to fishers in easily understandable ways were the key factors that enabled an agreement to be reached. Unexpected income (tiger puffer for coastal fishers and blackmouth angler for offshore fishers) also provided economic support and softened the economic blow. This was decisive, but nobody could have predicted this degree of luck. The Sandfish in this case lived up to its name of "the God Fish."

References

Hirota M (2011) A study on the Sea cucumber distribution and its dynamics. Jpn J Fish Econ 55(1):129–148 (in Japanese)

Kawauchi-machi Fisheries Cooperative Association (1982–2009) Business Report

Makino M (2011) A theory of the management options for sea cucumber fisheries in Japan: a simulation analysis with management objectives and area characteristics. Jpn J Fish Econ 55(1):149–165 (in Japanese)

Marine Eco-Label Japan (2010) http://www.melj.jp/index.cfm. Accessed 30 June 2010

Ministry of Agriculture, Forestry and Fisheries (1982–2009) Annual statistics of fisheries and aquaculture production. Nourin-toukei-kyokai, Tokyo

Ministry of Finance (2005–2009) Japan exports and imports: commodity by country. Japan Tariff Association, Tokyo

Suenaga S (2008) Sandfish resource co-management in Akita Prefecture, Japan. In: Townsend R, Shotton R, Uchida H (eds) Case studies in fisheries self-governance (FAO Fisheries Technical Paper No. 504). Food and Agriculture Organization of the United Nations, Rome, pp 191–200

Tamaki Y (2004) Fisheries management of sandfish in Akita Prefecture. In: Proceedings of International Institute of Fisheries Economics and Trade (IIFET) 2004, Tokyo, Japan

Tomiyama M, Yanagibashi S (2004) Effect of temperature, age-class and growth on induction of sand eel aestivation in Ise Bay, Central Japan. Fish Oceanogr 13:81–90

Tomiyama M, Komatsu T, Makino M (2008) Sand eel fisheries governance in Ise Bay, Japan. In: Townsend R, Shotton R, Uchida H (eds) Case studies in fisheries self-governance (FAO Fisheries Technical Paper No. 504). Food and Agriculture Organization of the United Nations, Rome, pp 201–210

Yamaguchi T (2007) History of coastal fisheries. Seizando Shoten, Tokyo (in Japanese)

Chapter 5
Fisheries Management in Offshore Areas

Abstract Two cases from offshore areas are presented in this chapter. The first case is snow crab management by bottom trawlers off *Kyoto* Prefecture. This fishery is semi-industrialized with a relatively small number of vessels. After the severe collapse of the snow crab resource in the late 1970s, the organization of Kyoto bottom trawlers implemented several autonomous measures, including setting up no-take zones, for which they were awarded Japan's first Marine Stewardship Council (MSC) certificate. The second case is industrial purse seiners in the northwestern Pacific. In this area, a species alternation phenomenon has been observed among sardine, anchovy, and mackerel, which follows an approximately 50-year cycle. The purse seiners' response to sardine population fluctuations in the 1980s resulted in the failure of the chub mackerel bloom in the late 1990s. In response, the national government and the organization of purse seiners adopted, in 2003, a Resource Recovery Plan to protect strong year classes of chub mackerel, which is now gradually showing a successful outcome.

5.1 Snow Crab Bottom Trawlers off Kyoto Prefecture

5.1.1 Background

Kyoto Prefecture is located in the central part of the main island (Fig. 1.1), and has about 320 km of coastline facing the Japan Sea. In offshore areas, two types of fisheries are in operation: purse seiners targeting sardine, jack mackerel, chub mackerel, etc., and bottom trawlers targeting snow crab, brown sole, deep-sea smelt, etc. This section describes the case of bottom trawlers.

As of 2009, there were 15 Kyoto-based bottom trawlers in operation. Their most important target species in terms of value was snow crab (*Chionoecetes opilio*). These are called "Pine Needle Crabs" in Japanese, due to the shape of their legs (Fig. 5.1). The subtle flavor and beautiful shape of the snow crabs make them very

M. Makino, *Fisheries Management in Japan: Its institutional features and case studies*, Fish & Fisheries Series 34, DOI 10.1007/978-94-007-1777-0_5,

Fig. 5.1 Snow crab (hard shell male)

popular in Japan, particularly as a winter specialty in northern Kyoto Prefecture, and many tourists visit the area to enjoy snow crab cuisine. Snow crabs landed in the Taiza area, branded as "Taiza Crabs," are especially renowned.

In the early 1960s, Kyoto bottom trawlers landed more than 500 ton of snow crab every year, but the yield had dramatically declined to less than 100 metric tons in the late 1970s. To deal with this situation, various management measures have been adopted, mostly since 1982. As a result, catch volume and value have both shown a gradual recovery. The Kyoto Bottom Trawlers' Union, which is a fisheries management organization (FMO) composed of local bottom trawlers, has played a central role in their management (Makino 2008). In 2008, they were awarded Japan's first Marine Stewardship Council (MSC) certificate.

Female snow crabs molt ten times in 5–6 years, and after the final molting (the 11th molting stage, shell size 7–8 cm), they mate and become primiparous females. The main spawning season of snow crab is September for primiparous females and March for multiparous females. With male crabs, there is some individual variability in the timing of the terminal molt (from the 11th to 16th molting stages, shell size 15 cm). It is thought that they take about 7–8 years to reach maturity. The molting season, for both males and females, is from September to October. Mating occurs at depths of 220–290 m, peaking at around 270 m (Yamasaki 1994; Kon et al. 2003).

5.1.2 Snow Crab Fishery

Local records show that hand-powered bottom net fishing started on the Kyoto Prefecture coast in the mid-twelfth century. Powered trawlers, which were introduced in this area in 1919, rapidly increased fishing capacity. This naturally led to conflicts between fisheries. Therefore, in 1922, the government imposed a license system for bottom trawl fishing, and established Bottom Trawl Prohibition Zones to keep bottom trawlers about 3 nm away from the coast. In 1944, the Kyoto Bottom Trawlers' Union

Fig. 5.2 Small-scale bottom trawlers operating at sea

was founded to represent the interests of bottom trawl fishery production in Kyoto. This organization, which is composed of local bottom trawlers and is therefore an FMO (Table 2.2), has played a central role in fisheries governance in this area until today.

There are 15 bottom trawlers based in Kyoto Prefecture, of which 13 are smaller than 15 gross tons. These are called Small-Scale Bottom Trawlers and are licensed by the prefectural governor (Fig. 5.2). The two vessels larger than 15 gross tons are categorized as Offshore Bottom Trawlers, licensed by the minister (Sect. 3.2.4). They harvest snow crab in winter at around 200–350 m depth, along the coastal edge of the Kyoto offshore area. The deeper areas are used by larger bottom trawlers with home ports in neighboring Prefectures. In other seasons, Kyoto bottom trawlers harvest brown sole (*Pleuronectes herzensteini*), sandfish (*Arctoscopus japonicus*), deep-sea smelt (*Clossanodon semifasciatus*), firefly squid (*Watasenia scintillans*), pointhead flounder (*Hippoglossoides pinetorum*), willowy flounder (*Tanakius kitaharai*), blackmouth angler (*Lophiomus setigerus*), etc. From June to August, bottom trawling is prohibited by ministerial ordinance and the Kyoto Prefecture Fishery Coordinating Regulations.

Snow crabs are commercially classified into three types: hard-shelled crabs (males more than 1 year after their last molt, which fetch high prices), soft-shelled crabs (males just before or after molting, with low prices), and females. In fiscal year 2009 (from April 2009 to March 2010), bottom trawlers in Kyoto Prefecture harvested 97 ton, worth JPY 315 million, of snow crab (Table 5.1). For bottom trawlers, snow crab is the most important target species in terms of value.

Table 5.1 Snow crab production in *Kyoto*

Fiscal year	Volume (ton)	Value (JPY million)	Unit price (JPY/kg)
1964	369	135	366
1965	294	146	496
1966	323	165	512
1967	273	132	483
1968	226	119	527
1969	288	167	581
1970	243	161	662
1971	266	189	710
1972	213	214	1,006
1973	100	122	1,218
1974	119	157	1,317
1975	113	134	1,189
1976	102	148	1,450
1977	105	144	1,369
1978	80	136	1,701
1979	76	134	1,769
1980	58	109	1,878
1981	137	168	1,228
1982	97	141	1,451
1983	86	166	1,928
1984	85	210	2,470
1985	69	205	2,976
1986	66	207	3,142
1987	84	243	2,894
1988	76	255	3,359
1989	118	284	2,405
1990	133	357	2,686
1991	101	320	3,169
1992	80	291	3,635
1993	99	310	3,128
1994	146	415	2,841
1995	158	449	2,841
1996	165	417	2,525
1997	138	355	2,573
1998	128	244	1,905
1999	195	340	1,744
2000	168	371	2,206
2001	169	423	2,504
2002	134	353	2,631
2003	138	352	2,549
2004	105	329	3,135
2005	120	281	2,339
2006	125	326	2,610
2007	112	318	2,855
2008	78	299	3,822
2009	97	315	3,241

Source: National Federation of Bottom Trawlers' Unions (2010)

Note: This table is based on fiscal years (April–March), so values are slightly different from Table 5.2, which is based on the fishing year

Table 5.2 TACs for snow crab for the national total, Western Japan Sea Stock, and *Kyoto* Prefecture

Fishing year	Snow crab TAC (ton)	Real landed volume (ton)	Quota for the western Japan Sea stock (ton)	Real landed volume (ton)	Quota for *Kyoto* Prefecture (ton)	Real landed volume (ton)
1997	4,815	4,333	2,100	2,492	–	–
1998	4,945	4,307	2,205	2,808	–	–
1999	5,469	4,461	2,426	2,942	–	–
2000	5,469	5,030	2,465	2,786	–	–
2001	5,469	4,900	2,670	3,186	–	–
2002	6,775	5,001	3,079	3,178	145	108
2003	6,605	5,635	3,295	3,410	207	111
2004	7,218	5,271	3,716	3,886	241	86
2005	7,453	5,224	4,087	3,652	265	99
2006	7,113	5,631	4,523	3,960	130	102
2007	7,224	5,587	4,580	4,107	132	95
2008	7,793	4,834	4,698	3,425	147	65
2009	6,423	–	3,629	–	91	–

Source: Japan Fisheries Information Service Center (2010)
Note: The fishing year for snow crab was from January to December until 2000, and from July to June from 2001. Quota allocation to *Kyoto* Prefecture started from the fishing year of 2002

5.1.3 Management

5.1.3.1 Formal Management Framework

The formal management frameworks for Kyoto bottom trawl fisheries are the fishery licenses, the Kyoto Prefecture Fishery Coordinating Regulations, and several ministerial ordinances and notifications. The total number of vessels engaged in these fisheries is regulated by licenses, i.e., Small-Scale Bottom Trawlers (less than 15 gross tons) and Offshore Bottom Trawlers (more than 15 gross tons), licensed by the prefectural governor and the minister of MAFF, respectively. The total number of licenses for Small-Scale Bottom Trawlers in Japan is also prescribed and allocated to each Prefecture by the minister of MAFF. Therefore, this kind of Governor-Licensed Fishery is called a Minister-Allocated Governor-Licensed Fishery. Their fishing season (from September 1st to May 31st) is officially prescribed by a ministerial notification and the Kyoto Prefecture Fishery Coordinating Regulations.

Depending on the target species, there is another formal framework for licensed fisheries. Governor-licensed fisheries targeting certain species which need to be managed uniformly throughout the nation are categorized as Specified Minister-Licensed Fisheries, and boat owners have to get an additional license from the minister of MAFF to harvest these species. Snow crab is listed in this category. Additional regulations prescribed by ministerial ordinance govern all Specified Minister-Licensed Fisheries. Also, as introduced in Sect. 2.4.1, a Total Allowable Catch (TAC) has been set for snow crab since 1997 in accordance with the United Nations Convention on the Law of the Sea (UNCLOS). Table 5.2 shows TAC values and real landed volumes since 1997. The snow crab TAC is divided into five stocks (the western Japan Sea, northern Japan Sea, Western Hokkaido Island, Sea of Okhotsk, and Northern Pacific), and the bottom trawlers in Kyoto Prefecture are using the western Japan Sea stock with neighboring Prefectures.

5.1.3.2 Autonomous Management Measures

In addition to the formal regulatory frameworks explained above, a wide range of regulations have been implemented to protect snow crab resources and produce greater value. The most important of these measures is Marine Protected Areas (MPAs). The MPAs for snow crab resources in Kyoto consists of two types: permanent marine reserves (no-take zones) and autonomous restraint of operations in spring and autumn.

The first type of MPA aims to provide snow crabs with perpetual sanctuaries from fishing pressure. Since 1983, local trawl fishers have successively introduced six no-take zones in critical habitats for snow crabs, such as reproduction areas, male habitats, female habitats, etc., based on scientific advice from the prefectural research station (Yamasaki 2002). Also, to ensure that trawl fishing pressures are completely excluded from these protected areas, concrete blocks 3 m in length on each side were placed on the sea floor at a density of 3.8/km^2. The cost of these concrete blocks was borne by the government. In 2009, the total area of the marine reserves was 67.8 km^2, corresponding to 4.4% of the total fishing ground area for snow crab fishing.

The second type of MPA involves restraint of operations based on autonomous agreements among bottom trawlers. As explained above, the fishing season for bottom trawlers is from September 1st to May 31st. On the other hand, the fishing season for snow crab is only in winter: from the beginning of November to nearly the end of March. Bottom trawl operations targeting other species in spring and autumn were therefore conducted at exactly the same depth as the snow crab's habitat, and as a result, a lot of snow crabs were caught as bycatch until the mid-1980s. When fishers captured snow crabs during these seasons, they could not be sold at market and had to be dumped back into the sea. Most young or soft-shelled crabs are killed by this treatment. According to Yamasaki and Kuwahara (1991), about 45–60% of the initial snow crab stocks in each year were unintentionally destroyed due to this mixed catching.

Soft-shelled crabs are destined to become hard-shelled crabs. The latter is more than 10 times the price of the former, so it is very important to protect young soft-shelled crabs. Therefore, the second type of MPA, the autonomous restraint of operations, aims to prevent bycatch of soft-shelled crabs during the off seasons of snow crab (spring and autumn). Based on agreements among bottom trawlers, operations at the snow crabs' habitat (200–350 m depth) are restrained on an autonomous basis. This period of restraint comes to an end when the snow crab season begins in winter.

Several other measures have also been implemented on an autonomous basis, such as additional shortening of fishing seasons, stricter minimal size setting, and improvements to equipment. The mesh size of trawling nets has been steadily enlarged by agreement among bottom trawlers. Since 2003, a new technology, the crab exclusion system, was installed in nets to prevent bycatch of snow crabs in spring and autumn. The maximum catch limits per single sortie are also set by agreement. Finally, since 2008, all landings of soft-shelled crabs are prohibited.

The result of these measures was that the resource status gradually recovered, starting in 1983. One thing worth pointing out is that the improvement in the average catch value per day is much greater than that in the average catch volume per day, meaning that the average value of each landed snow crab is improving, probably because of the shift in fishing pressure from soft-shelled crabs to hard-shelled crabs, as well as increased size resulting from the other measures. Statistics show that the unit price considerably has improved from 1,200 to 1,800 JPY/kg in the 1970s to around 3,000 JPY/kg in recent years (Table 5.1).

Imported snow crabs from Canada, Russia, North Korea, etc., fetch much lower prices than their Japanese equivalents. So, to differentiate the Kyoto products from imported ones, the bottom trawlers in Kyoto attach a plastic tag to landed snow crabs by which consumers can readily identify their place of origin. This producer tag for Japanese snow crab is now commonly used in the western part of the Sea of Japan.

5.1.3.3 FMO Structure

The principal decision maker in this autonomous framework is the Kyoto Bottom Trawlers' Union, which was founded in 1944. All the bottom trawlers in Kyoto belong to this union. Through this union, the trawlers know each other and meet on a regular basis, and have built up mutual trust for generations. There is also a federation of trawlers' unions, the National Federation of Bottom Trawlers' Unions, which coordinates cross-jurisdictional matters among Prefectures. As explained in Sect. 2.4.1, this union is the main body for snow crab TAC implementation.

The local research institute, the Fisheries Technology Department of Kyoto Prefectural Agriculture, Forestry and Fisheries Technology Center, has also played a crucial role as an advisor for local bottom trawlers' decision making. For example, the construction of marine reserves was initially proposed by a researcher at the institute. The size and locations of the six reserves were decided by the bottom trawlers based on scientific information presented by the institute. The research results estimating the irrational dissipation of soft-shelled crab resources from bycatches in spring and autumn led to the restraint on operations during these periods.

5.2 Chub Mackerel Purse Seiners off the Pacific Coast

5.2.1 Background

5.2.1.1 Species Alternation Among Sardine, Anchovy, and Mackerel

It is well known that the catch composition in the northwestern Pacific changes periodically due to natural fluctuations in resource abundance. It appears that the Japanese sardine (*Sardinops melanostictus*), anchovy (*Engraulis japonicus*),

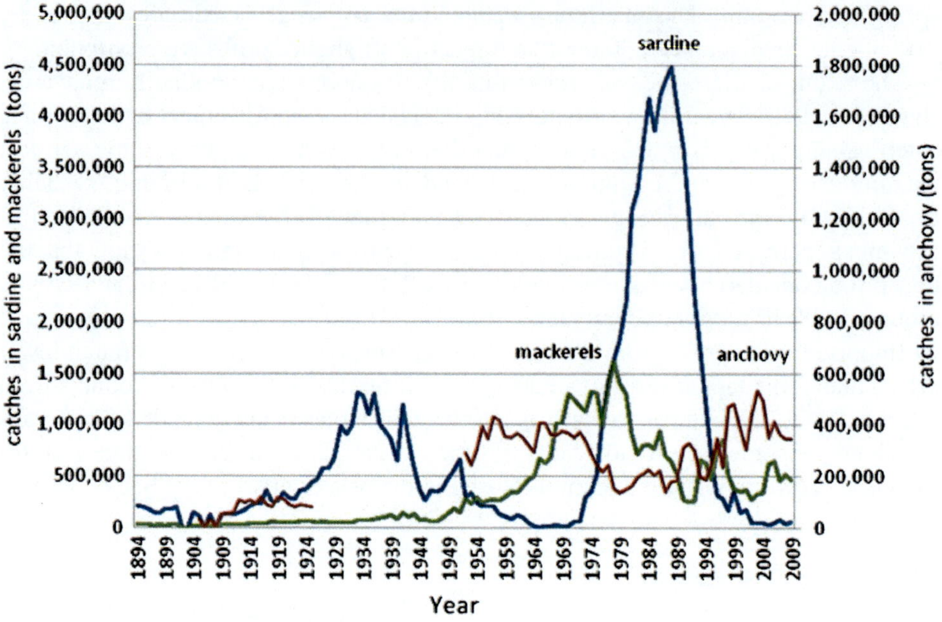

Fig. 5.3 Changes in Japanese catches of sardine, anchovy, and mackerel (Source: MAFF 1895–2010) Note: In the MAFF statistics report, no statistics on anchovy values were recorded between 1894–1904 and 1926–1951. The sum of anchovy and sardine productions were recorded as "sardines" during these periods

and chub mackerel and spotted chub mackerel (*Scomber japonicus* and *Scomber australasicus*) show a species alternation phenomenon with an approximately 40–50-year cycle (Yatsu et al. 2003, Fig. 5.3).

There is growing evidence that environmental regime shifts, abrupt transitions of environmental variables from one set of condition to alternative state(s), can influence plagic fish productivities (Lluch-Belda et al. 1989; Beamish 1995; Wada and Jacobson 1998; Tanaka 2003; Yatsu et al. 2005). For example, the collapse of the Japanese sardine in the late 1980s (Fig. 5.3) coincided with a period of large meanders in the Kuroshio Current and positive sea surface temperature anomalies in the Kuroshio Extension South Area (Nakata et al. 1994; Noto and Yasuda 2003). These phenomena can be understood as a kind of natural fluctuation in the North Pacific marine ecosystem. It should thus be kept as it is, and not flattened by human intervention.

Assuming that species alternations between the three species actually take place, the mackerel catch should have increased since around the end of the 1990s. Since we have not observed any increase, it appears that there has been a disruption of the fish alternation phenomenon (ecosystem dynamics) in the northwestern Pacific. One reason for this disruption was the purse seiners' response to sardine fluctuation in the 1980s, as explained in this section.

5.2.1.2 Development of Purse Seine Fishing in Japan

Japanese purse seine fishing started in the central part of Japan in the early seventeenth century. The net was about 600 m in length, and operated with 9 boats crewed by more than 50 fishers (Oh-unabara 1980). In the late nineteenth and early twentieth century, a new purse seine technology was imported from the United States and modified to fit the Japanese operational style. It was called "improved purse seine fishery," and is the direct ancestor of today's purse seiners in Japan. The operation size of the improved purse seine fisheries was much smaller than the original Japanese purse seine in the Edo Era, and required only 26 crew and a net about 200 m long. The development of the domestic cotton industry at the same period considerably cut the price of fishnets, and gave impetus to the speed of proliferation of the improved purse seiners.

There was then another major technological step forward. Until this time, all the fishing boats in Japan were powered by oars or sails. The first engine-powered fishing boat was launched in 1906. With financial support from the government as a national policy, this new technology took over in the 1920s. Engine-powered purse seiners also rapidly developed all around Japan, and their fishing grounds finally expanded into the areas around the Korean Peninsula. In 1941, about 77% of the total sardine catch (970,000 ton) in Japan was harvested by purse seiners. Also, at this time, mackerel became another important target species for purse seiners.

During WWII, a lot of fishers as well as purse seine boats were lost in the fighting, and the total production of Japanese purse seiners fell dramatically from 845,000 ton in 1941 to 189,000 ton in 1945. To cope with domestic food shortages, the recovery of purse seine fishing was very quickly achieved with various legal and financial inducements from the government. In 1952, the Fisheries Agency promulgated a new regulatory rule for purse seine fisheries based on the Fisheries Law of 1949, and established the Minister License system for purse seiners. This basic management framework still applies.

5.2.2 *Interrelationship Between Species Alternation and Purse Seine Fishing*

The main fishery type which catches sardine, anchovy, and mackerels in the northwestern Pacific is Large- and Medium-Scale Purse Seine Fishery (Sect. 3.2.5). This is one of the most industrialized and largest fisheries sectors in Japan. There are eight jurisdictional sea areas defined for the management of this fishery. The Northern Pacific area, which is the Pacific offshore area from Chiba Prefecture to Hokkaido Prefecture, is the most productive area for purse seiners. This section deals with the Large- and Medium-Scale Purse Seiners in this area.

In the Northern Pacific area, large- and medium-scale purse seine fishing is conducted by the operation unit which is usually composed of four vessels: one main ship (the net boat), one search ship, and two transport ships. The total number of fishers in one operation unit is 40–50.

Table 5.3 Catch volume and real values by large- and medium-scale purse seiners operating in the Northern Pacific Area

Year	Volume (ton)	Value (JPY million)
1972	480,387	18,953
1973	479,593	16,727
1974	562,845	30,805
1975	696,629	23,621
1976	712,123	39,334
1977	1,086,497	38,565
1978	1,150,100	30,120
1979	1,298,960	47,172
1980	1,366,317	51,942
1981	1,665,762	62,067
1982	1,742,208	59,559
1983	1,926,724	64,709
1984	2,096,599	57,321
1985	1,887,797	56,027
1986	2,397,061	50,122
1987	1,997,745	49,784
1988	1,997,970	47,798
1989	1,521,763	40,479
1990	1,233,897	32,758
1991	1,060,051	33,772
1992	937,657	29,567
1993	914,933	24,200
1994	512,139	26,100
1995	376,875	30,156
1996	416,383	25,246
1997	670,554	34,759
1998	576,700	37,327
1999	508,707	40,403
2000	370,336	33,056
2001	445,942	33,176
2002	312,896	27,660
2003	349,399	20,096
2004	325,388	26,043
2005	413,099	24,432
2006	506,819	33,601
2007	378,534	38,717
2008	352,709	39,618
2009	280,398	26,471

Source: Association of Purse Seine Fisheries in the North Pacific area (2008)

5.2.2.1 Sardine Bloom and Decline in the 1980s

In the mid 1980s, about 72 operation units in the Northern Pacific area harvested around 2 million tons every year. The catch value peaked at JPY 64.7 billion in 1983. On the other hand, the catch volume peaked at 2.4 million tons in 1986 (Table 5.3).

The majority of the catch in the 1980s was sardine, and the price of sardine around 1986 was about half of that in 1983. In this period, fishers say, the huge numbers of sardines in the fishing ground sometimes caused purse seine nets to break.

The business situation between 1983 and around 1986 was as follows: there were many sardines in the Northern Pacific area but the market price was very low, and total revenue was continuously decreasing. So, the purse seiners' natural response was to increase the number of operation units to catch more sardine. However, the total number of nets (operation units) was regulated by licenses from the minister, so they could not invest in building new operation units.

Therefore, the purse seiners' actual response was to construct large-scale transport ships (over 300-ton type). According to the vessel registration statistics, gross tonnages of this class of transport ship increased more than 2.5 times in the 8 years after the catch value peak in 1983 (MAFF 1983, 1991). According to our interviews with local purse seiners, the actual depreciation period for transport ships was about 3–4 years. As a result, they competed to build new ships in this period to avoid the high tax rates charged on their high income. Japan was also experiencing an economic bubble at this time, so the banks were happy to lend money to purse seiners to build new ships.

In 1988–1989, the oceanographic conditions around Japan suddenly changed, and sardine numbers went into sharp decline in 1989–1990. The total catch volumes in 1990, 1991, and 1992 were only 51%, 44%, and 39% of that in 1986, respectively. In the end, too much borrowed capital was invested in equipment and too few sardines were left in the sea.

5.2.2.2 Effects on Mackerel Resources

After the sardines had gone, the purse seiners still had to pay back their debts. In the 1990s, to escape bankruptcy, they directed their fishing efforts toward mackerel. Note that bankruptcy is a very serious thing in Japanese culture, the social equivalent of a death sentence. Every year, many Japanese commit suicide because of bankruptcy. I have personally known several cases. My view is that the purse seiners really had no alternative at that time.

As a result, heavy fishing pressure was put on the Northern Pacific stock of chub mackerel by purse seiners in the 1990s. In this period, 80–90% of the mackerel caught were immature fish weighing less than 300 g (Fisheries Agency and Fisheries Research Agency 2009). We had year classes of chub mackerel in 1992 and 1996 that were strong enough to be called a mackerel bloom. However, because of overintense fishing pressure from the purse seiners, most of these fish were harvested before they could reproduce, and this prevented species alternation. From around 1992, the number of operation units had been gradually reduced by cutting the number of licenses and financial support from the government, but the situation could not be overcome.

To sum up, resource fluctuation of sardines in the 1980s and fishing fleet dynamics corresponding to that fluctuation are closely related to the current resource level of chub mackerel. Although there were enough anchovy in the 1990s, their price was much lower than that of small mackerel. These market conditions prevented anchovy from being an economic substitute for sardine or chub mackerel. Makino and Mitani (2010) discussed several policy interventions that should have been adopted around this period, and numerically simulated their possible effects on total catch and resource biomass of chub mackerel.

5.2.3 Management

5.2.3.1 Management Organizations

Because the Large- and Medium-Scale Purse Seiners operating in the Northern Pacific area are the Minister-Licensed Fisheries, and their home ports extend for about 1,000 km along the pacific coastline, the main authority in the management is the central government (Fisheries Agency). However, there are several levels of FMOs composed of Large- and Medium-Scale Purse Seiners who also play a positive role. For example, purse seiners operating offshore of each Prefecture have set up prefectural organizations of purse seiners. Similarly, prefectural organizations in the Northern Pacific area have set up the North Pacific Federation of Purse Seiners. This is also another main body for coordination and management in the Northern Pacific area. Similar federations are organized in other areas as well. As the umbrella organization, the National Federation of Purse Seiners comprises all the local federations of purse seiners and the associations of distant-water (high sea) purse seiners.

5.2.3.2 Management Measures

Formal Measures

Based on the Ministerial Ordinance on the Licensing and Policing of Designated Fisheries of 1965, eight jurisdictional sea areas are defined for the management of the Large- and Medium-Scale Purse Seiners. Also, based on this ordinance, purse seiners using a net ship bigger than 15 gross tons in the Northern Pacific area are categorized as Middle- and Large-Scale Purse Seine Fisheries (Table 1.3). The total number of operation units is regulated by licenses from the minister of MAFF. Each license lasts 5 years. Their fishing grounds and fishing seasons are prescribed in detail according to the above ordinance and several other official rules set by the central government. In addition, as the limitations or restrictions on the license, more detailed limits on fishing grounds, equipment, and seasons are applied to each operation unit.

As the output control measure, TAC has been adopted since 1997 (Table 5.4). In 2003, a Resource Recovery Plan for chub mackerel was adopted, as explained later.

Table 5.4 TACs for sardine and mackerel (chub mackerel and spotted chub mackerel)

	Sardine		Mackerel	
Fishing year	TAC (1,000 ton)	Real landed volume (1,000 ton)	TAC (1,000 ton)	Real landed volume (1,000 ton)
1997	720	268	700	726
1998	520	154	700	466
1999	400	336	780	345
2000	380	137	780	333
2001	380	162	770	334
2002	342	46	693	256
2003	100	45	512	329
2004	70	40	419	311
2005	60	19	575	612
2006	60	51	655	611
2007	60	71	746	404
2008	52	31	765	471
2009	61	49	548	–

Source: Japan Fisheries Information Service Center (2010)
Note: Until 2005, the fishing year for sardine was from January to December. Since 2006, the fishing year for mackerel has been from January to December and from July to June

Autonomous Measures

Organizations of purse seiners play an important role in coordinating with coastal fisheries. One example can be found in the Hachinohe area of Aomori Prefecture. This area is famous as a good fishing ground for Japanese common squid (*Todarodes pacificus*) and mackerels. In 1965, purse seiners started operating in this area and came into serious conflict with coastal squid fishers. In 1966 and 1976, the North Pacific Federation of Purse Seiners and the associations of coastal squid fishers (mainly composed of pole and line fishers) signed a mutual agreement on the operational limits to purse seiners. Although the conflicts have not been completely resolved, this framework has been developed with the continuous administrative support of the Fisheries Agency, and now the mutual agreement includes a maximum squid catch volume for purse seiners. Similar examples can be found in Chiba Prefecture, where coastal pole and line fishers and purse seiners have upheld a mutual agreement since around 1962 on the mackerel catch at the mouth of the Tone River. Also, as explained in Sect. 2.4.1, federations of purse seiners play an important role in implementing TACs for sardine and mackerel.

Resource Recovery Plan and Its Results

At the end of the 1990s, heated discussions broke out about overfishing by purse seiners of the strong year classes of chub mackerel that occurred in 1992 and 1996. Based on the views exchanged at these deliberations, the North Pacific Federation

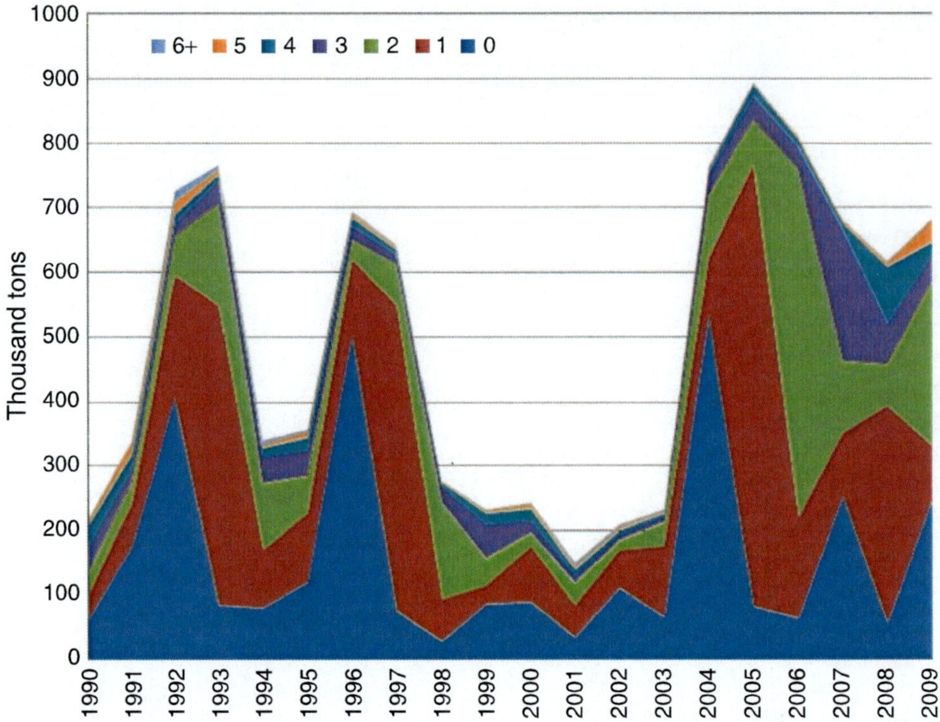

Fig. 5.4 The estimated biomass of each year class from 1989 to 2009 (Data from Fisheries Agency and Fisheries Research Agency 2010)

of Purse Seiners, the National Federation of Purse Seiners, members of the Pacific Wide-Area Fisheries Coordinating Committees (Table 2.2), government officers, and fisheries researchers compiled a Resource Recovery Plan for the Pacific stock of chub mackerel. It came into force in 2003 and is effective until 2011.

Adaptive protection of strong year classes is especially important for widely fluctuating resources such as chub mackerel. The objective of the plan is thus to protect strong year classes by cutting fishing pressure on them, mainly from the North Pacific Large- and Medium-Scale Purse Seine Fishery. The recovery target was set as a spawning stock biomass (SSB) of over 180,000 ton, corresponding to a sixfold increase from the SSB in 2003.

Based on the Resource Recovery Plan, the North Pacific Federation of Purse Seiners effected their fishing pressure reduction plan by cutting their number of operating days (days at sea) and total number of operating units. This reduction plan is a form of adaptive strategy, with several reduction scenarios made contingent on the appearance of strong year classes. If a strong year class appears, fishing pressure must be cut by about 25–30%.

The next year, in 2004, a strong year class occurred, and in 2005, probably as a result, the total landed volume of mackerel exceeded the TAC (Table 5.4). One reason for this was insufficient reduction of fishing pressure on the strong year class that had occurred in the previous year. This represented a serious failure of the Japanese TAC enforcement system, jointly implemented by the national government and the purse seiners' organization (Sect. 2.4.1). If the TAC in 2005 had been more strictly enforced, today's resource levels could have been better.

Two years later, in 2007, we had a middle-strength year class of chub mackerel. This time, the North Pacific Federation of Purse Seiners adopted an additional individual-vessel quota system over the last several months of the fishing year (July–June) to ensure that total landings never exceeded the TAC and thus to effectively protect strong year classes. Since then, the total catch has not exceeded the TAC. In 2009, we had a relatively strong year class: the purse seiners instigated 31 no-operation days in the fishing year of 2009, corresponding to a 38% cut in fishing pressure.

As a result of these measures, the Pacific stock of chub mackerel is now showing a gradual recovery, as can be seen in Fig. 5.4. According to the latest report submitted from the Fisheries Research Agency to the Pacific Wide-Area Fisheries Coordinating Committee, the SSB in 2009 was estimated to be about 230,000 ton, which considerably exceeds the recovery target of the Resource Recovery Plan of 180,000 ton. The next step should be to increase the SSB to 450,000 ton, above which, based on the historical data, stable yearly production can be expected.

References

Association of Purse Seine Fishery in the North Pacific area (2008) Catch Report

Beamish R (1995) The need to understand the relationship between climate and the dynamics of fish populations. Canadian Special Publication of Fisheries and Aquatic Sciences 121:1–2

Fisheries Agency and Fisheries Research Agency (2010) Stock assessment report on the Pacific stock of chub mackerel. http://abchan.job.affrc.go.jp/. Accessed 30 June 2010 (in Japanese)

Japan Fisheries Information Service Center (2010) Catch report of TAC managemed species. http://www.jafic.or.jp/tac/index.html. Accessed 30 June 2010 (in Japanese)

Kon T, Adachi T, Suzuki Y (2003) Distribution of snow crab, Chionoecetes spp. larvae off Wakasa Bay in the Sea of Japan. Fish Sci 69:1109–1115

Lluch-Belda D, Crawford RJM, Kawasaki T, McCall AD, Parrish RH (1989) World-wide fluctuations of sardne and anchovy stocks: the regime problem. S Afr J Mar Sci 8:195–205

Makino M (2008) Marine protected areas for the snow crab bottom fishery off Kyoto Prefecture, Japan. In: Townsend R, Shotton R, Uchida H (eds) Case studies in fisheries self-governance (FAO Fisheries Technical Paper No. 504). Food and Agriculture Organization of the United Nations, Rome, pp 211–220

Makino M, Mitani T (2010) Fisheries management under species alternation: a case of the Pacific purse seiner off Japan. In: Proceeding for International Institute of Fisheries Economics and Trade (IIFET) 2010, Montpellier, France

Ministry of Agriculture Forestry and Fisheries (1895–2010) Annual report of fisheries and aquaculture productions. Nourintoukei-kyokai, Tokyo

Ministry of Agriculture Forestry and Fisheries (1983, 1991) Statistics on Fisheries Vessels. Nourintoukei-kyokai, Tokyo

Nakata K, Harada A, Matsukawa Y (1994) Variations in food abundance for Japanese sardine larvae related to the Kuroshio meander. Fish Oceanogr 3:39–49

National Federation of Bottom Trawlers' Union (2010) Report of snow crab catch at the Japan Sea in fiscal year of 2009

Noto M, Yasuda I (2003) Empirical biomass model for the Japanese sardine, *Sardinops melanostictus,* with surface temperature in the Kuroshio Extension. Fish Oceanogr 12:1–9

Oh-unabara K (1980) History of purse seine fishery. In: National Federation of Purse Seiners (ed) Ten year history. National Federation of Purse Seiners, Tokyo, pp 3–210

Tanaka E (2003) A method for estimating dynamics of carrying capacity using time series of stock and recruitment. Fish Sci 69:677–686

Wada T, Jacobson LD (1998) Regimes and stock-recruitment relationships in Japanese sardine (*Sardinops melanostictus*), 1951–1995. Can J Fish Aquat Sci 55:2455–2463

Yamasaki A (1994) Studies related to resource management: ecological characteristics of the Tanner crab, Chionoecetes opilio. Research Report of Kyoto Governmental Oceanography Center 4:1–53

Yamasaki A (2002) Establishment of preserved area for snow crab *Chionoecetes opilio* and consequent recovery of the crab resources. Fish Sci 68:1699–1702

Yamasaki A, Kuwahara A (1991) Operations and issues in bottom trawl fishery off Kyoto Prefecture. Suisan no Kenkyu 55:109–114 (In Japanese)

Yatsu A, Nagasawa K, Wada T (2003) Decadal changes in abundance of dominant pelagic fishes and squids in the Northwestern Pacific Ocean since the 1970s and implications to fisheries management. Trans Am Fish Soc Symp 38:675–684

Yatsu A, Watanabe T, Ishida M, Sugisaki H, Jacobson LD (2005) Environmental effects on recruitment and productivity of Japanese sardine *Sardinops melanostictus* and chub mackerel *Scomber japonicus* with recommendations for management. Fish Oceanogr 14(4):236–278

Chapter 6
Institutional Relationship Between Japanese Fisheries Management and the Ecosystem Approach

Abstract Based on the Ecosystem Approach of the Convention on Biological Diversity (CBD) and case studies presented in Chaps. 4 and 5, this chapter examines the institutional advantages of and necessary measures for developing Japanese fisheries management into ecosystem-based management. Japanese fisheries management has advantages that include a decentralized management system, use of local and scientific knowledge, multi-scale and interlinked management frameworks, etc. On the other hand, measures need to be adopted to cover the following areas: adoption of ecosystem perspectives, wider stakeholder involvement, an ecosystem monitoring system, and appropriate use of indicators and protected areas. The conclusions derived in this chapter are later used to assess the ecosystem-based management applied to the Shiretoko World Natural Heritage area to be discussed in Chap. 8.

6.1 The Convention on Biological Diversity and the Ecosystem Approach

The Convention on Biological Diversity (CBD) is the first global agreement on the conservation of biological diversity and the sustainable use of its components. It was adopted in June 1992 at the United Nations Conference on Environment and Development (UNCED) held in Rio de Janeiro, along with the United Nations Framework Convention on Climate Change (UNFCC), the Rio Declaration on the Environment and Development, the Declaration of Forest Principles, Agenda 21, etc.

The CBD set out a commitment for maintaining the world's ecological underpinnings alongside economic development. It established three main goals: (1) conservation of biodiversity, (2) sustainable use of the components of biodiversity, and (3) sharing the benefits arising from the commercial and other utilization of genetic resources in a fair and equitable way (CBD 1992). In order to deliver these goals, the Ecosystem Approach was adopted as the primary framework for action under the Convention (decision II/8).

M. Makino, *Fisheries Management in Japan: Its institutional features and case studies*, Fish & Fisheries Series 34, DOI 10.1007/978-94-007-1777-0_6,

Table 6.1 Principles and operational guidance of the ecosystem approach (From CBD 2000)

Principle 1	The objectives of management of land, water, and living resources are a matter of societal choice.
Principle 2	Management should be decentralized to the lowest appropriate level.
Principle 3	Ecosystem managers should consider the effects (actual or potential) of their activities on adjacent and other ecosystems.
Principle 4	Recognizing potential gains from management, there is usually a need to understand and manage the ecosystem in an economic context. Any such ecosystem-management program should: (a) Reduce those market distortions that adversely affect biological diversity; (b) Align incentives to promote biodiversity conservation and sustainable use; (c) Internalize costs and benefits in the given ecosystem to the extent feasible.
Principle 5	Conservation of ecosystem structure and functioning, in order to maintain ecosystem services, should be a priority target of the ecosystem approach.
Principle 6	Ecosystem must be managed within the limits of their functioning.
Principle 7	Ecosystem approach should be undertaken at the appropriate spatial and temporal scales.
Principle 8	Recognizing the varying temporal scales and lag effects that characterize ecosystem processes, objectives for ecosystem management should be set for long term.
Principle 9	Management must recognize the change is inevitable.
Principle 10	The ecosystem approach should seek the appropriate balance between, and integration of, conservation and use of biological diversity.
Principle 11	The ecosystem approach should consider all forms of relevant information, including scientific and indigenous, and local knowledge, innovations, and practices.
Principle 12	The ecosystem approach should involve all relevant sectors of society and scientific disciplines.
Guidance 1	Focus on the functional relationships and processes within ecosystems.
Guidance 2	Enhance benefit sharing.
Guidance 3	Use adaptive management practices.
Guidance 4	Carry out management actions at the scale appropriate for the issue being addressed, with decentralization to lowest level, as appropriate.
Guidance 5	Ensure intersectoral cooperation.

The Ecosystem Approach is based on the application of appropriate scientific methodologies focused on levels of biological organization which encompass the essential processes, functions, and interactions among organisms and their environment. It also recognizes that humans, with their cultural diversity, are an integral component of ecosystems. Decision V/6 endorsed the description of the 12 principles of the Ecosystem Approach and the 5 operational guidelines for its application (Table 6.1).

This chapter focuses on the Ecosystem Approach of the CBD, and describes its conceptual background. Institutional characteristics of Japanese fisheries management are then examined from the viewpoint of the Ecosystem Approach. The objective of this analysis is to define the advantages and shortcomings of Japanese fisheries management, and to derive logical foundations for the policy responses necessary to achieve marine ecosystem conservation.

6.2 The Ecosystem Approach as an Alternative to Conventional Fisheries and Environmental Policy

6.2.1 Changes in Problem Perception

For the last several decades, human beings have been aware of their adverse impacts on ecosystems, and have been taking measures to combat these problems. In 1972, the first global conference on environmental issues, the United Nations Conference on the Human Environment, was held in Stockholm. One of the core issues of the conference was environmental destruction and pollution caused by industrialization and urbanization. In the process of the discussions, however, sharp divisions became apparent between the developed countries who were focusing on the environmental degradation caused by developmental activities, and the developing countries who wished to eradicate domestic poverty and starvation through industrialization and development. With regard to this situation, the World Commission on Environment and Development, also known as the Brundtland Commission, published a report on their 4-year discussions, "Our Common Future," in 1987. They presented the idea of "sustainable development," meaning development that meets the needs of the present without compromising the ability of future generations to meet their own needs. In other words, the report proposed a new concept of development that was not antagonistic to environmental conservation, but essential to satisfy human needs and improve the quality of the environment and human life. The point is that development must be based on the efficient and environmentally responsible use of all of society's scarce resources—natural, human, and economic.

During these decades, concurrently, there has been a shift in the problems facing the world's environment and a new awareness that the issues to be tackled are not limited to local, specific pollution or destruction, involving only visible stakeholders (polluters and victims), but are broader in scope, including global environmental deterioration affecting a wide range of stakeholders, such as global warming and loss of biodiversity (Matsushita 2002). This paradigm shift in environmental issues led inevitably to a fundamental change in environmental policy ideas, that is, from regulatory policies such as setting environmental standards or the construction of preservation areas, to integrated and holistic social strategies aiming to ensure wise use and conservation of society's scarce resources. The CBD and other global environmental conventions and declarations adopted at UNCED in 1992 were an offshoot of this trend.

When it comes to wildlife issues, the traditional approaches found, for example, in the Convention on International Trade in Endangered Species of Wild Fauna and Flora (CITES) of 1973 or the Ramsar Convention on Wetlands of 1975, are basically single-species or sectoral approaches that do not take into account interactions among ecosystem components, including human activities. As a consequence of international trends in environmental policy concepts as described above, the concept of ecosystem management (Christensen et al. 1996) has attracted wide attention as a way to redeem these shortcomings.

Ecosystem management is defined as a management philosophy that focuses on desired states rather than system outputs and which recognizes the need to protect or restore critical ecological components, functions, and structures to sustain resources in perpetuity (Cortner et al. 1994). It aims at (1) maintaining viable populations of all native species in situ; (2) representing within protected areas all native ecosystem types across their natural range; (3) maintaining evolutionary and ecological processes; (4) managing over periods of time of sufficient duration to maintain the evolutionary potential of species and ecosystems; and (5) accommodating human use and occupancy within these constraints (Garcia et al. 2003).

The Ecosystem Approach of the CBD has many aspects in common with ecosystem management. It is not a set of guidelines tailored to the management needs of various ecosystem types. In fact, the Ecosystem Approach under the CBD is a framework for holistic decision-making and action which links biological, social, and economic information and aims to achieve a socially acceptable balance between nature conservation priorities and the use and sharing of the benefits of resources (Smith and Maltby 2003). Therefore, in addition to the bottom line for managing ecosystems, such as Principles 3, 5, 6, 7, and 8, it highlights socio-economic factors. It firmly recognizes people as an integral component of ecosystems and puts them at the center of management, and engages the widest range of sectoral interests via a participatory approach and decentralized institutional framework (Principles 1, 2, and 12). In addition, it uses adaptive measures to deal with uncertainty (Principle 9), and stresses the economic context, such as through appropriate incentive measures (Principles 4, 10). In doing this, all kinds of information from scientific, administrative, local, and indigenous levels are considered (Principle 11), and social investment to empower these actors is strongly encouraged. In summary, the Ecosystem Approach is not a replacement, but an extension of conventional species/area-specific management practices, and can be understood as a new environmental policy framework to achieve ecosystem management.

6.2.2 Ecosystem Approach for Marine Ecosystem Conservation and Fisheries

Although the development of the ecosystem concept in marine conservation is still in its infancy, there are several global conventions, agreements, and mandates which specifically deal with its application to marine ecosystems. The first global convention that applied the ecosystem concept to ocean management was the 1980 Convention on the Conservation of Antarctic Marine Living Resources (CCAMLR). Chapter 17 of Agenda 21 also deals with marine ecosystems and recognizes integrated coastal zone management (ICZM) to be a very effective tool. ICZM is defined as "a mechanism for bringing together a multiplicity of users, stakeholders, and decision-makers in the coastal zone to secure more effective ecosystem management whilst achieving economic development and intra- and inter-generational equity through the application of sustainability principles." (Ramsar Convention 2002). Although there are several similar terminologies, such as integrated marine and coastal area management (IMCAM), integrated coastal management (ICM), or integrated coastal area management (ICAM), they are all based on a similar idea. In summary, the ICZM and other approaches such as IMCAM, ICAM, or ICM, are social strategies for achieving marine ecosystem management and are recognized to be the most effective tools for achieving the aims of the CBD (AID Environment 2004).

Other major international documents relating to marine ecosystem conservation include the United Nations Convention on the Law of the Sea (UNCLOS) of 1982, the FAO Code of Conduct for Responsible Fisheries and Kyoto Declaration at the Conference on the Sustainable Contribution of Fisheries to Food Security of 1995, the Jakarta Mandate on Marine and Coastal Biological Diversity in 1995, and the Johannesburg Declaration on Sustainable Development at the World Summit on Sustainable Development (WSSD) in 2002, to list a few (FAO 2003; Hanna 2003).

In the fisheries context, the ecosystem perspective became an important pillar of the management framework after the Reykjavik Declaration on Responsible Fisheries in Marine Ecosystems in October 2001. The Reykjavik Declaration calls for, inter alia: (1) immediate adoption of management plans with incentives for the sustainable use of ecosystems, (2) strengthening of governance, (3) prevention of adverse effects of non-fisheries activities on marine ecosystems and fisheries, (4) advances on a scientific basis for incorporating ecosystem considerations into management (including the precautionary approach), (5) monitoring interactions between fisheries and aquaculture, (6) strengthening international collaboration, (7) technology transfer, (8) removal of trade distortions, (9) collection of information on management regimes, and (10) development of guidelines.

There have been a number of terms developed to describe fisheries and ecosystem conservation. Some have focused more on the natural science ecosystem components, while others have stressed a more holistic and integrated (interdisciplinary) interpretation. In response to an international call for assistance to clarify what is meant by the "ecosystem approach" in the fisheries context, the FAO organized Technical Consultations on this issue in Reykjavik in September 2002, and published guidelines on the ecosystem approach to fisheries (FAO 2003). The FAO proposed the following definition. It is aligned with the more general ecosystem approach idea, but seeks to strike a pragmatic balance by focusing on aspects that are within the ability of fisheries management bodies to implement, while recognizing the fisheries sectors' responsibility to collaborate in a broader multisectoral application of the ecosystem approach (De Young et al. 2008).

> An ecosystem approach to fisheries (EAF) strives to balance diverse societal objectives by taking account of the knowledge and uncertainties of biotic, abiotic and human components of ecosystems and their interactions and applying an integrated approach to fisheries within ecologically meaningful boundaries.

Table 6.2 Typology of the ecosystem approach to management

Ecosystem-based management component	I. Traditional single-factor management	II. Sectoral management in an ecosystem context	III. Integrated management in an ecosystem context
Species	Considers only the factor or species being used	Considers prey, dependent predators and food supply, and impacts on the ecosystem	Considers impacts of other activities on the status of the species being used and across the ecosystem
Physical habitats	Only considered if a surrogate for population parameters	Considers productive capacity and impacts of activity on the habitat	Accommodates spatial needs and habitat impacts of other activities
Environmental conditions	Not considered	Considers productivity regime and forcing	Considers direct and indirect effects
Biodiversity	Not considered	Considers impacts on species not being used directly	Considers status of communities and resilience of the community/system

Source: Fluharty et al. (2010)

Another very similar concept is the Ecosystem-based Fisheries Management (EBFM). The U.S. National Research Council defined EBFM as follows.

> An approach that takes major ecosystem components and services—both structural and functional—into account in management fisheries...It values habitat, embraces a multispecies perspective, and is committed to understanding ecosystem processes... its goal is to rebuild and sustain populations, species, biological communities and marine ecosystems at high levels of productivity and biological diversity so as not to jeopardize a wide range of goods and services from marine ecosystems, while providing food, revenue and recreation for humans.

Fisheries activities are managed under the EAF or EBFM framework. In this sense, it is an extension of the traditional single-sector approach. On the other hand, even the traditional single-sector approach (fisheries management or single species management) does take a considerable amount of ecosystem information into account. Fluharty et al. (2010) pointed out that the more factors and species that are taken into account in management decisions, the greater the progress toward EA/EBFM (Table 6.2).

6.3 Assessment of Japanese Fisheries Management from the Viewpoint of the Ecosystem Approach of the CBD

In this section, an analytical framework for the ecosystem approach, which consists of six interconnected themes (CBD 2003), is applied to assess Japanese fisheries management (Makino 2005). These are (1) provision of environmental goods and services (what is being managed within ecosystems and for what purpose?); (2) building a consensus (who will undertake the task of management?); (3) providing incentives for management (what are the incentives for management?); (4) balancing the conservation and use of biotic resources (how can different management objectives be reconciled and integrated?); (5) cross-scale integration (how best to integrate management across multiple scales of interaction and response?); and (6) building adaptive capacity (how best to develop the capacity to initiate, learn from, and thereby sustain activities?).

6.3.1 Provision of Environmental Goods and Services (Principle 5)

Marine fisheries are an industry which utilizes marine ecosystem services (a provisioning service). Consequently, the conservation, use, and management of fisheries resources should take place in an ecosystem context. These activities have potential consequences in terms of changes in the structure and functioning of the marine ecosystem of which the fisheries resources are part. This can affect the production of goods and services, either positively or negatively. Whether such impacts arise, and with what effects, depends greatly on the particular structure of the ecosystem concerned, the nature of the linkages among its components, and the resulting processes and functions.

Of course, the main focus of Japanese conventional fisheries management has been on target species, especially highly valued ones, and it has not paid much attention to the ecosystem context per se. The reason simply relates to the principal aim of the current Fisheries Law that was enacted shortly after WWII: that is, to develop fisheries productivity to cope with domestic food shortages and to improve the economic status of small-scale fishing people actually engaged in fishing operations (Sect. 2.3). However, in the Fisheries Basic Policy Plan of 1999, it is clearly stated that ecosystem conservation is a prerequisite for fisheries management. Also, Article 2 of the Basic Act on Fisheries Policy of 2001 (Sect. 2.4.3) recognizes that fisheries resources are a component of the marine ecosystem, and require conservation. Ecosystem management has thus become an increasingly important policy task in Japanese fisheries administration.

To achieve marine ecosystem management, first of all, scientific understanding should be facilitated with regard to the relationship and interactions between fisheries target species and ecosystems, the impact of fishing operations on the ecosystem, and effective measures for its conservation. For example, Matsuda and Abrams' multi-trophic food web model (2006) shows that simple yield or profit maximization policy may lead to the extinction of a significant fraction of the species. Zhou et al. (2010) discussed the nature of selective processes in fisheries operations (species selection, stock selection, size selection, sex selection, season selection, species selection), and suggested the concept of "Balanced Exploitation" to mitigate the negative effects of the selection. Therefore, to conserve the ecosystem structure of the system, there will be a need for institutions to coordinate decision making between fisheries targeting prey species and those targeting predator species, based on monitoring results of the ecosystem structure in concern.

6.3.2 *Building Consensus (Principles 1, 11, and 12)*

Human society is diverse in the kind and manner of relationships that different groups have with the natural world, each viewing the world around them in different ways and emphasizing their own economic, cultural, and societal interests and needs. For example, as shown in Fig. 1.2, fisheries are a food-providing industry and have a particularly important social role in many countries, including Japan. Hence, determination of the methods of use or conservation objectives of marine ecosystems is inevitably a sociological issue.

For example, Jamieson and Zhang (2005) conducted an international comparison of marine ecosystem conservation in north Pacific countries (Canada, China, Japan, Korea, Russia, and the USA). Their study concluded that there are clear differences between the West and the East. China, Japan, and Korea have greater coastal populations and a longer history of full exploitation and development. They face the challenges of minimizing existing impact, rebuilding depleted stocks, lessening impact from land runoff, etc. On the other hand, in Canada, Russia, and the USA, human coastal populations and development are at a much lower level: their challenges are to maintain wild, pristine habitat and appropriately economically active communities.

Therefore, the objective of marine ecosystem conservation should be a "societal choice (Principle 1)." In this regard, marine resource users (fishers) and local communities are especially important stakeholders, since they live on the resource and can more directly affect its future. Their rights and interests have to be appropriately recognized and incorporated into the management planning process. At the same time, the involvement of all relevant stakeholders and technical expertise in planning and carrying out joint activities, as well as sharing management resources, is essential for effective management (Principle 12).

In Japan, fisheries have long been the principal industry in many coastal communities, and hence, fisheries coordination has had a lot in common with local social needs for the use of marine ecosystems. However, there are almost no other interests taken into account in the decision-making process for this fisheries coordination. For example, recreational anglers' interests are relatively important around urban areas such as *Tokyo* Bay or *Osaka* Bay, but there are very few channels for the official inclusion of recreational anglers' interests in fisheries management regimes (Makino 2002). In successful cases, where coordination between local fishers and recreational anglers is going very smoothly, the owners of angling charter boats are often members of the local FCA.

Other than fisheries and recreational angling, there is a broad range of users in marine ecosystems. This means there are varieties of interests related to marine ecosystems, such as nonuse value for citizens, or an afflux of land-oriented pollutants or nutrients through material circulation with coastal areas. Also, the availability of the fishery resources as a bequest to future generations, or their potential to provide new goods such as pharmaceuticals can have value as options. In order to take all these into account, the viewpoint of watershed management is important, and a coordinating system should be devised which transparently reflects all the relevant stakeholders' interests. This system can be established as an extension of the current fisheries coordinating organization (Table 2.2) by incorporating various stakeholders into the organization, or separately established with the fisheries industry as a constituent.

There are specific marine ecosystem functions and structures in each geographical and seasonal condition. They are actually infinite in variety. The local fishers and their organizations have a lot of explicit and tacit knowledge of the local area which has accumulated for generations. As explained in the first half of this book, local fishers are the core of fishery management in Japan. Therefore, the local knowledge accumulated by local fishers should be utilized in ecosystem management (Principle 11). Daily catch data are also an important source of ecosystem information. However, information provided by local fishers alone is not sufficient to achieve conservation of ecosystem functions and structures. Therefore, as a first step, priorities for additional information or data necessary for ecosystem conservation need to be identified from a scientific point of view. A role-sharing scheme should next be devised for data collection and monitoring among local fishers, the government, and other citizens.

6.3.3 *Providing Incentives for Management (Principle 4)*

Marine ecosystems provide economically valuable goods and services, thus predicating the need to understand and manage ecosystems in an economic context (Principle 4). In that sense, organizing local fishers as resource managers provides an incentive to manage the resource effectively so as to cut associated costs or gain additional benefits from enhanced ecosystem services. Indeed, in Japan, there have been traditional activities which can be appraised from the ecosystem point of view. As explained using the cases presented in the next chapter, fishers have voluntarily planted sea grasses in coastal areas, and established coastal nursery grounds and marine protected areas. Others have afforested upstream hills (*Uo-Tsuki-Rin*, fish conservation forests). These activities have a long history, and are worthy of remark and research. However, they are motivated mainly by their potential effects on the target species rather than the larger ecosystem services. In other words, these activities can be understood as "fishery ground conservation" and take place in an economic context.

There are several advantageous features of the Japanese institution with regard to promoting ecosystem management in-line with Principle 4. Legally protected rights/licenses mean a high level of protection of their interests (Dolsak and Ostrom 2003). This high degree of security can then lead to the economic motive for sustainable use of the area which they are coordinating, since the incentive to conserve the long-term ecosystem health of that area may outweigh, for example, individual transferable quotas (ITQs) that are defined for each target resource per se. This difference in the characteristics of interests, that is, area-based or target resource-based, may be more influential in uncertain conditions, where rights such as ITQs are redefined as new information becomes available, and decreases the security of rights, which then reduces the economic motivation for long-term decision making. Therefore, in Japan, if a proposed ecosystem conservation plan can demonstrate an economic benefit for fishers, they will be willing to play a positive role by contributing their knowledge and abilities.

In addition to these incentives at the production stage, distribution stage-oriented incentives such as fisheries certificates (e.g., Marine Eco-Label Japan, or Marine Stewardship Council, etc.) are now being taken up in Japan, and are expected to act as additional measures to reduce adverse effects on ecosystems. In sum, Japanese fishery management has the potential to expand its scope to ecosystem conservation in-line with Principle 4.

However, the high rate of fishing gear depreciation and very high uncertainty in fishery income will raise the discount rate, and the future term considered by each fisher during the decision-making process tends to be relatively short. Also, as mentioned in Sect. 3.1, 59.5% of Japanese fishers have no one in mind to take over their business, and this situation inevitably leads to shorter perspectives when it comes to decision making. Negative externalities from other fisheries operations as well as land-oriented pollutants and recreational fishing may complicate the situation. To resolve this timescale inconsistency in decision making, the key is to make reference to long-term performance indicators and maintain adaptive attitudes in decision making.

6.3.4 Balancing Conservation and the Use of Biotic Resources (Principles 6 and 10)

Principle 10, keeping an appropriate balance between conservation and use, is the very same concept governing Resource Management-Type Fisheries (*Shigen Kanri-gata Gyogyo*) as presented in Sect. 2.3.3 and Chaps. 4 and 5. But its main scope is, again, limited to economically valuable species. Economic incentives are critical and must be fully utilized for efficient conservation, but the scope of management should be broadened as far as possible to include ecosystem structures and functions. Likewise, limits to their functioning (Principle 6) are fundamentally acknowledged in fisheries management as far as the target resources are concerned. The TACs or TAEs system is a formal institution for keeping fisheries pressures within certain limits (Sect. 2.4). The task now is to incorporate ecosystem perspectives into TAC/TAE formulating protocols.

A system of marine protected areas (MPAs) is attracting international attention as an option for ensuring the conservation of marine ecosystems. Within a system of MPAs, a range of measures can be applied along a continuum from ecosystems that are strictly protected, through mixed resource-use systems, of which ecosystem conservation and sustainable use are both part, to areas that have been wholly utilized by human activities, including fisheries.

It is important to note that the term "MPAs" does not necessarily mean "no-take zones." No-take zones are just one form of the MPA system. In fact, they can range from small closed areas or harvest refugia designated to protect a specific resource or habitat type, to extensive multiple-use MPA areas that integrate the management of many species, habitats, and uses in a single, comprehensive plan (Agardy et al. 2003). Misunderstandings of the meaning of MPA often lead to negative reactions, especially if local communities feel that something is being taken away from them to preserve endangered species or ecosystems. As Sanchirico and Wilen (2001) implied by their meta-population bioeconomic model, according to the biological nature and market price of the resource and the level of its overexploitation, an appropriately designed system of MPAs can improve both fishery income and ecosystem health. Therefore, a system of MPAs is not antagonistic to fisheries operations, but can be understood as an ecosystem-based resource enhancement system which enables multiple and responsible use of ecosystem services, including fishery operations. The case of snow crab fishery management using MPAs (Sect. 5.1) is an example that conforms to this concept. Definitions, issues, and case examples of MPAs in Japan are given in more detail in Chap. 7.

The limits of ecosystems are not static, but may vary across sites, through time, and in response to past circumstances and events. There is considerable uncertainty and ignorance about the actual limits (thresholds for change) in different ecosystems. The impact of MPAs is also uncertain and difficult to predict. The externalities raised from other sectors add more complexity. Management, whether for conservation or planned sustainable resource use, should thus be adaptive and flexible.

6.3.5 *Cross-Scale Integration (Principles 2, 3, 7, and 8)*

Principle 2 states that natural resource management is best carried out at the level of the resource production system. This is a variation of the subsidiarity principle, which states that higher-tier authorities should not assume functions that can be carried out more appropriately by lower-tier associations; or, to put it another way, problems are best resolved at the level of the organization at which they occur. This principle is in-line with the current trend toward increasing devolution of responsibility for natural resource management to local institutions, on the grounds of greater efficiency, effectiveness, and equity (CBD 2003).

As described in Chap. 2, the basic concept of Japanese fisheries management is that of resource management by the resource users themselves. The challenge is, therefore, to incorporate ecosystem viewpoints into the management system on an ecologically meaningful scale of area and time.

Scale, however, is a serious question in ecosystem conservation (Principles 7 and 8). How the components are perceived to be arranged spatially depends partly on the scale of observation. There is no single level of organization at which to understand and best manage ecosystem functioning. Each level—genetic, population, species, community, or landscape—is important or irrelevant, depending on the nature and scale of the problem being addressed and on the perspective and aims of the managers.

Likewise with time: on one time scale (e.g., monthly or annually) a component or process may appear to exhibit constant periodicity; on another, longer or shorter time scale, the temporal dynamics may appear to be episodic or chaotic (unpredictable). In addition, ecosystems are not closed systems (Principle 3). They are largely open, connected to other systems through the flow of energy, matter, and information, and the movement of organisms. To cope with these problems of scale, effective management institutions need to be devised at multiple and interconnected levels (McGinnis and Ostrom 1996).

In Japanese fisheries management, there are various levels of management organizations, from the local community level to the national level, as shown in Table 2.2. To transcend jurisdictional boundaries, FMOs are organized by members from several FCAs or several prefectures according to the biological nature of the target species (see the cases in Chaps. 4 and 5). The Japanese management system therefore has the potential to cope with geological-scale problems. However, again, the current system focuses chiefly on target species. For example, in regard to Principle 3, nearly 80 species of artificial fish seeds are released along Japanese coasts to enhance fisheries resources, but the potential effects on adjacent and other ecosystems are not sufficiently investigated. Another issue relating to space is territorial disputes. As many other countries do, Japan has territorial disputes with neighboring countries. Marine ecosystems naturally straddle these man-made lines and boundaries. An example of such case is presented from the Shiretoko World Natural Heritage area in Chap. 9.

As for timescale problems, fishers tend to follow shorter-term incentives than the timescale of ecosystem evolution. Political interests also have a tendency to be a few years in length. Hence, some institutional arrangements should be delivered to guarantee long-term conservation objectives. To resolve this timescale inconsistency in decision making, performance indicators that summarize data on complex environmental issues to indicate the overall status and trends of marine ecosystems would be a useful tool (Rice and Rochet 2005; Perry et al. 2010). Competitive fisheries operations based on economic incentives should be utilized as long as these long-term performance indicators are within scientifically and transparently determined ranges. The TAC/TAE system in Japan can be understood as one form of ceiling limit in current fisheries management, but they are based on the single-species management perspective, and ecosystem perspectives such as interspecies effects are not included at this moment.

6.3.6 Building Adaptive Capacity (Principle 9)

Change in ecosystems is both natural and inevitable. Ecosystems change, including with regard to species composition, population abundance, and human–resource interactions.

Hence, management should adapt to the changes, as discussed in the fisheries management under species alternation phenomena (Sect. 5.2). Building the flexibility and capacity to adapt to new situations is critical for the success of the management.

In Japanese fisheries management, in which local fishers are the principal decision makers, any management decisions can be changed in response to information gained via fishing operations. In other words, there is considerable flexibility in management decision making. In addition, the local FCAs or FMOs offices are used as capacity-building and information centers for management. In this sense, Japanese fisheries management is potentially capable of adaptive decision making.

As presented in Sect. 5.1, one of the most successful cases of adaptive resource management in Japan is the snow crab fishery by bottom trawlers off Kyoto Prefecture. Two kinds of management measures were in place there: autonomous restraint in fishery operations and no-take zones constructed at public expense. In both cases, a small-scale trial was implemented before full expansion. Based on the information from fishery operations during the trial period, organizations of local fishers decided on expansion of each measure, and considerably reduced the economic risk (Makino 2004).

A future task is to develop infrastructures for data collection and monitoring, in which local fishers and their organizations should play an important role, and to incorporate ecosystem perspectives into these adaptive decision-making processes.

6.4 Discussion

This chapter pointed out that, from the viewpoint of the Ecosystem Approach of the CBD, Japanese fisheries management has its own institutional advantages, such as a decentralized management system, adaptive decision-making process, use of local and scientific knowledge, multi-scale and interlinked management, and promotion of sustainable resource use in an economic context.

To be able to develop the Japanese fisheries management framework into ecosystem-based management, the key issues to be tackled are summarized as follows.

First, as repeated many times, a fundamental weakness is the lack of clear concern for the impact of fisheries operations on marine ecosystems. More progress needs to be made in scientific understanding of ecosystems' structures, functions, and processes, and consistent institutional reforms should be carried out.

However, when imposing reforms, full consideration must be given to the institutional contexts and background of Japanese fisheries management. As discussed in Sect. 2.2, we cannot transfer an institutional design that worked well at one place to another place and expect to repeat the success. The characteristics of fishery rights/licenses and coordination systems in Japan would logically be transferable to a different ecosystem conservation framework from, for example, that of Iceland or New Zealand, where individual transferable quotas (ITQs) are the central management tool. Therefore, what is required is a careful examination of the appropriate

Table 6.3 Institutional advantages and necessary policy measures taken by Japanese fisheries management assessed from the viewpoint of the ecosystem approach of the CBD

Institutional advantages

- Decentralized management systems by local resource users.
- Use of local and scientific knowledge for management.
- Multi-scale and interlinked coordinating organizations.
- Adaptive management process based on daily fishing operations.
- Promotion of sustainable resource use in an economic context.

Necessary policy measures

- Ecosystem perspectives: progress in scientific knowledge should be facilitated. Careful examination of the appropriate nature of fishing rights/licenses and deliberate discussion of the role of the fishery industry in marine ecosystem-based management are required. The formulating protocols of TAC and TAE can also include ecosystem perspectives.
- Stakeholder involvement: new institutions should be set up to allow a wide range of stakeholders to be involved in transparent decision-making processes. The viewpoint of watershed management is also important.
- Data collection and monitoring: identification of priorities in ecologically necessary data, and role sharing in data collection and monitoring should be established.
- Indicators: development and reference to long-term indicators should be promoted. Fisheries should be operated within the allowable ranges of the indicators.
- Use of MPA systems: MPAs are not a synonym for no-take zones, but can be understood as an ecosystem-based resource enhancement system which enables multiple and responsible use of ecosystem services, including fishery operations. Economically and ecologically meaningful MPA systems can be devised and should be installed where necessary.

nature of the existing fishing rights/licenses, and purposeful discussions on the role of the fisheries industry in marine ecosystem conservation. The formulating protocols for TACs and TAEs could also include ecosystem perspectives.

Second, the scope of coordination or stakeholder participation is now limited to the fisheries sector only. No other marine ecosystem users are included in the decision-making or coordinating process. An appropriate system corresponding to material circulations within a watershed should be established, along with the necessary legal framework. What also needs to be seriously considered is a transparent decision-making procedure which reconciles and integrates a wide range of management objectives.

Third, adaptive management is needed to deal with uncertainty. The first step toward adaptive management is to identify priorities for necessary ecosystem information, then to determine the roles that fishers can take on, and develop supplementary data gathering and monitoring systems to fill the gap. Also, measurable indicators of the overall status and long-term trends of ecosystems need to be developed and continuously referred to.

Fourth, marine protected area (MPA) systems can be an effective tool for marine ecosystem conservation. Note that "MPA" is not a synonym for "no-take zone." Economically and ecologically meaningful MPA systems can be devised, and should be adopted where necessary.

Table 6.3 summarizes the institutional advantages and necessary policy measures for Japanese fisheries management.

Several issues in the Japanese institutional framework merit recognition and discussion. As pointed out in Chaps. 2 and 3, there are numerous structural defects in Japanese fisheries management. For example, autonomous decision making by local fishers can sometimes lead to inflexibility or reclusiveness. Vested interests may be overprotected or egalitarianist pressure may prevent the efficient use of ecosystem services. Local organizations may also prove to be unwilling to introduce new technologies, thereby retarding technical progress. A decentralized approach is not a versatile prescription. In addition, coordination by local fishing people is vital to the current institutional framework, but it inevitably becomes very complex and locally specific. Sometimes fishing people cannot play their expected roles as coordinators.

By the same token, the advantages listed in Table 6.3 do not hold in every area in Japan. Table 6.3 is just a general description of Japanese institutional potentialities. At successful sites, there are unique and site-specific conditions which enable the full realization of institutional potential. For example, in the cases of the fishery management activities presented in Chaps. 4, 5, and 8, very active information exchanges have continued among local fishing people, the local research station, and prefectural government officials. This is a form of cross-sectoral communication that is characteristic of Japanese fisheries management. In addition, long cross-generational acquaintances among fishers operating in the area are one of the most important factors in building relationships of mutual trust. As Rustagi et al. (2010) pointed out, resource users who have higher proportion of "conditional cooperators" and costly duties were more likely to succeed in the management of common resources. Therefore, in addition to the institutional framework analyzed in this chapter, the capacities, functions, and relationships of the people concerned should constitute the very core of marine ecosystem conservation activities.

References

Agardy T, Bridgewater P, Crosby MP, Day J, Dayton PK, Kenchington R, Laffoley D, McConney P, Murray PA, Parks JE, Peau L (2003) Dangerous targets? Unresolved issues and ideological clashes around marine protected areas. Aquat Conserv MarFreshw Ecosyst 13(4):353–67

AID Environment (2004) Integrated Marine and Coastal Area Management (IMCAM) approaches for implementing the Convention on Biologica Diversity (CBD technical series no. 14). Secretariat of the Convention on Biological Diversity, Montreal

Christensen NL, Bartuska AM, Brown JH, Carpenter S, D'Antonio C, Francis R, Franklin J, MacMahon JA, Noss RF, Parsons DJ, Peterson CH, Turner MG, Woodmansee RG (1996) The report of the ecological society of America committee on the scientific basis for ecosystem management. Ecol Appl 6:665–691

Convention on Biological Diversity (1992) Convention on Biological Diversity. http://www.cbd.int/convention/text. Accessed 30 June 2010

Convention on Biological Diversity (2000) Decisions adopted by the conference of the parties to the convention on biological diversity at its fifth meeting. UNEP/CBD/COP/5/23

Convention on Biological Diversity (2003) Review of the principles of the ecosystem approach and suggestions for refinement: a framework for discussion. UNEP/CBD/EM-EA/1/3

Cortner HJ, Shannon MA, Wallace MG, Burke S, Moote MA (1994) Institutional barriers and incentives for ecosystem management (Issue paper 16). University of Arizona, Water Resources Research Centre, Tucson

De Young C, Charles T, Hjort A (2008) Human dimensions of the ecosystem approach to fisheries: an overview of context, concepts, tools and methods (FAO fisheries technical paper 489). Food and Agriculture Organization of the United Nations, Rome

Dolsak N, Ostrom E (2003) The challenges of the commons. In: Dolsak N, Ostrom E (eds) The commons in the new millennium. MIT Press, Cambridge, pp 3–34

Fluharty D, Harvey C, Jamieson G, Jin X, Livingston P, Makino M, Radchenko V, Zhang CI (2010) Developing an ecosystem based approach for ocean management in the PICES region. In: Jamieson G, Livingston P, Zhang CI (eds) Report of working group 19 on ecosystem-based management science and its application to the North Pacific (PICES scientific report 37). North Pacific Marine Science Organization (PICES), British Columbia, pp 8–14

Food and Agriculture Organization of the United Nations (2003) The ecosystem approach to fisheries (FAO technical guidelines for responsible fisheries 4, Suppl. 2). Food and Agriculture Organization of the United Nations, Rome

Garcia SM, Zerbi A, Aliaume C, Do Chi T, Lasserre G (2003) The ecosystem approach to fisheries (FAO fisheries technical paper 443). Food and Agricultural Organization of the United Nations, Rome

Hanna S (2003) Transition in the American fishing commons: management problems and institutional design challenges. In: Dolsak N, Ostrom E (eds) The commons in the new millennium. MIT Press, Cambridge, pp 61–76

Jamieson G, Zhang CI (2005) Report of the study group on ecosystem-based management science and its application to the North Pacific (PICES scientific report 29). North Pacific Marine Science Organization (PICES), British Columbia

Makino M (2002) Legal characteristics of fishery rights, and pleasure fishing: case of the Ieshima affair. J Reg Fish 42(2):25–42 (in Japanese)

Makino M (2004) Using real options to estimate a benefit of the adaptive resource management. In: Proceeding for International Institute of Fisheries Economics and Trade (IIFET) 2001, Tokyo, Japan

Makino M (2005) The ecosystem approach of the CBD and fisheries management in Japan. Glob Environ Res 9:95–104

Matsuda H, Abrams PA (2006) Maximal yields from multi-species fisheries systems: rules for systems with multiple trophic levels. Ecol Appl 16(1):225–237

Matsushita K (2002) Environmental governance. Iwanami Shoten, Tokyo (in Japanese)

McGinnis M, Ostrom E (1996) Design principles for local and global commons. In: Young O (ed.) The international political economy and international institutions, vol 2. Edward Elgar, Cheltenham, pp 465–493

Perry I, Livingston P, Fulton B (2010) Ecosystem indicators. In: Jamieson G, Livingston P, Zhang CI (eds) Report of working group 19 on ecosystem-based management science and its application to the North Pacific (PICES Scientific Report 37). North Pacific Marine Science Organization (PICES), British Columbia, pp 83–90

Ramsar Convention (2002) resolution VIII/4, Appendix 1

Rice JC, Rochet M-J (2005) A framework for selecting a suite of indicators for fisheries management. ICES J Mar Sci 62:516–527

Rustagi D, Engel S, Kosfeld M (2010) Conditional cooperation and costly monitoring explain success in forest commons management. Science 330:961–965

Sanchirico JN, Wilen JE (2001) A bioeconomic model of marine reserve creation. J Environ Econ Manage 42:257–276

Smith RD, Maltby E (2003) Using the ecosystem approach to implement the convention on biological diversity: key issues and case studies (Ecosystem management series No. 2). International Union for Conservation of Nature (IUCN), Cambridge

Zhou S, Smith ADM, Punt AE, Richardson AJ, Gibbs M, Fulton EA, Pascoe S, Bulman C, Bayliss P, Sainsbury K (2010) Ecosystem-based fisheries management requires a change to the selective fishing philosophy. Proc Natl Acad Sci 107(21):9485–9489

Chapter 7
Marine Protected Areas

Abstract There are two types of marine-protected areas (MPAs) in Japan: Legal MPAs (LMPAs) and Autonomous MPAs (AMPAs). The former MPAs are established directly based on the prescriptions of legislative laws and set by the government. The latter MPAs are more site specific and responsive to the local socioecological conditions, and are established based on local initiatives. Fishers play a core role in their planning and implementation processes. Four cases of Autonomous MPAs presented in this chapter show the diversity of their objectives, participants, and implementation processes. After discussing the objectives of MPAs, the advantages and weak points of AMPAs are summarized. As other initiatives for ecosystem conservation by local fishing people in Japan, concepts of Sato-umi (the seas as part of one's homeland or community) and Uo-tsuki-rin (fish conservation forests) are briefly described.

7.1 Introduction

Marine-protected areas (MPAs) have been attracting increasing attention as a management tool for ecosystem conservation as well as fisheries resource conservation. For example, the plan for implementation of the World Summit on Sustainable Development (WSSD) at Johannesburg, 2002, includes the use of MPAs as an instrument that contributes to the conservation and management of oceans (WSSD 2002; Millennium Ecosystem Assessment 2005). The international community has committed to developing representative networks of MPAs by 2012 at WSSD and CBD COP 7. In addition, CBD COP 8, at Curitiba, Brazil, set the 2010 target to be "at least 10% of each of the world's ecological regions effectively conserved (CBD 2006)."

M. Makino, *Fisheries Management in Japan: Its institutional features and case studies*,
Fish & Fisheries Series 34, DOI 10.1007/978-94-007-1777-0_7,

In accordance with these international trends, the Japanese government stated the following in the Basic Plan on Ocean Policy (March 2008).

> As a means of ensuring biodiversity and to realize the sustainable use of fishery resources, the government should, in accordance with the Convention on Biological Diversity and other international agreements, clarify how to establish marine protected areas in Japan under coordination between related ministries and appropriately promote the establishment thereof.

There are no agreed single definitions of MPAs in the international community. Two of the most often referred definitions are those of "marine and coastal protected areas" by the Convention on Biological Diversity and of "protected areas" by the International Union for Conservation of Nature (IUCN).

> CBD's definition: (Marine and coastal protected area) means any defined area within or adjacent to the marine environment, together with its overlying waters and associated flora, fauna, and historical and cultural features, which has been reserved by legislation or other effective means, including custom, with the effect that its marine and/or coastal biodiversity enjoys a higher level of protection than its surroundings (CBD 2004).

> IUCN's definition: (A protected area is) a clearly defined geographical space, recognized, dedicated, and managed, through legal or other effective means, to achieve the long-term conservation of nature with associated ecosystem services and cultural values (Dudley 2008).

At this moment (December 2010), there are no authorized definitions of MPAs in Japan. This section adopts the legal definition of marine and coastal protected areas according to the Convention on Biological Diversity. The rationale for this is that (1) CBD is an official convention with legal binding force, and Japan is a contracting party; (2) the only official document mentioning MPAs in Japan is the Basic Plan on Ocean Policy, which also refers to the CBD in the same sentence.

7.2 Categories of MPAs in Japan

There is a wide range of MPAs, from types that try to exclude all human influence to ones that aim to achieve sustainable use by spatial management. The most well-known categorization by IUCN classifies protected areas into the following seven types (Table 7.1).

In Japan, MPAs can be categorized into two broad groups: the Legal MPAs (LMPAs) and the Autonomous MPAs (AMPAs), which are then divided into six and two subgroups, respectively (Makino 2010).

7.2.1 Legal MPAs (LMPAs)

The first category is Legal MPAs, which are established directly based on a legal framework. In Japan, there are at least seven types of Legal MPAs, as summarized in Table 7.2. The first four types of MPAs, i.e., Nature Conservation Area, Natural Park, Natural Coast Conservation Area, and Wildlife Protection Area, are administered mainly by the Ministry of the Environment. The other three types of MPAs are administered by the Fisheries Agency. These three types of MPAs are mainly for

Table 7.1 IUCN categories of protected areas (Source: Dudley 2008)

Type	Name	Features
Ia	Strict nature reserve	Native ecosystem in which human visitations, use, and impact are strictly controlled
Ib	Wilderness area	Unmodified or slightly modified areas which are protected and managed so as to preserve their natural state
II	National park	Large natural or near-natural areas providing a foundation for environmentally and culturally compatible spiritual, scientific, educational, recreational, and visitor opportunities
III	National monument of note	A specific natural monument with generally quite a small area. Often has high visitor value
IV	Habitat/species management area	Areas to protect particular species or habitats
V	Protected landscape/seascape	Areas where the interaction of people and nature over time has produced an area of distinct character of ecological, biological, cultural, and scenic value
VI	Protected area with sustainable use of natural resources	Areas for conserving ecosystems and habitats, together with associated cultural values and traditional natural resource management systems

conserving fisheries-related species or environment. Therefore, as discussed in detail in the last chapter, these MPAs have a lot of potential roles as a part of the system of MPAs, but not enough to achieve the holistic marine ecosystem-conservation. Also note that, in addition to these seven types of MPAs, there are other marine areas protected by the Ministry of Land, Infrastructure, Transport and Tourism, based on the Ports Law of 1950 or the Coasts Law of 1956. These areas are mainly for protecting physical structure of the marine environments.

Central ministries and local government have broad authority for planning, setting, and implementing Legal MPAs. However, consensus-building with local stakeholders is still very important for them. For example, there are many fisheries communities within Natural Parks, since local people engaged in fishing activities at almost every point along the Japanese coast for hundreds of years before the establishment of the National Park system, and their activities remain protected by fisheries rights/licenses. According to the 11th Census of Fisheries conducted in 2003, 44% of fisheries areas are located inside the Natural Parks. Consensus-building with local fishers is thus a prerequisite if the government plans to set up Legal MPAs in Japan.

7.2.2 Autonomous MPAs (AMPAs)

The second category of marine protected areas is Autonomous MPAs, based on local initiatives. Although Legal MPAs are set widely and regulate human impact throughout the areas, the distinct characteristic of Autonomous MPAs is that they are set on an issue-specific basis.

Table 7.2 Legal marine protected areas in Japan

Name of MPA	Legal basis	Objective	Contents	Estimated area
Nature conservation area	Nature Conservation Law of 1952	Protection of special natural environment	Regulation of catch of designated species, man-made constructions, digging, land reclamation, etc.	About 128 ha
Natural park	Natural Parks Law of 1957	Protection of outstanding marine scenery	Regulation of catch of designated species, man-made construction, digging, land reclamation, etc.	About 1.7 million ha
Natural seashore conservation area	Special Law on the Seto Inland Sea Environmental Protection of 1973	Preservation of natural status and public use	Regulation of man-made constructions, digging, land reclamation, etc.	91 areas
Wildlife protection area	Wildlife Protection and Appropriate Hunting Law of 2002	Conservation of wild animals	Regulation of catch of designated species, man-made constructions, digging, land reclamation, etc.	About 4,900 ha
Common fisheries right area	Fisheries Law of 1949	Conservation of fisheries resources and its habitat	FCA-based rules for resource use and conservation, exclusion of non-FCA members, etc.	(All along the Japanese coastline)
Protected waters	Fisheries Resource Protection Law of 1951	Protection of fisheries resources for spawning, nurseries, etc.	Regulation of land reclamation, use of specified equipment, catch of designated species, etc.	About 2,950 ha
Coastal fisheries development area, designated area	Marine Fisheries Development Promotion Law of 1971	Conservation of fisheries resources and habitats	Regulation of environmental modification and development	

Note: Area information provided by the Ministry of the Environment, Japan

There are numerous Autonomous MPAs along the Japanese coastline. According to the Fisheries Agency's survey conducted on Fisheries Cooperative Associations (FCAs) in 2006, 57.1% of FCAs were implementing some kind of spatial conservation measures for seagrass beds, tidal lands, or coral reefs (Fisheries Agency 2007a). Yagi et al. (2010) conducted a more comprehensive survey on both Legal and Autonomous MPAs from November 2009 to April 2010, and reported more than 1,161 MPAs, of which 1,003 had been autonomously adopted.

These Autonomous MPAs can be then grouped into two types: (1) resource management MPAs which aim to enhance specific target species for fisheries, and (2) ecosystem-based management MPAs with a broader perspective such as maintenance/restoration of marine ecosystems or endangered species conservation. Typical cases of the former type of Autonomous MPAs have already been presented in this book, such as sea cucumber (Sect. 4.2), sand eel (Sect. 4.3), and snow crab (Sect. 5.1). The next section therefore covers four examples of the latter type.

The historical records of Autonomous MPAs for the purpose of ecosystem-based management by local people can be traced back several hundred years or more. These activities typically focus on the primary productivity of marine ecosystems, such as seagrass beds, tidal lands, coral reefs, and forests in the upper streams. Although these activities do not directly target the enhancement of fisheries resources, they are a part of fisheries management and cannot be looked at in isolation. In contemporary phraseology, these activities are a part of the responsibilities borne by local fishers that come with the fisheries rights/licenses as ecosystem services users.

7.3 Cases of Autonomous MPAs in Japan

Unlike Legal MPAs, Autonomous MPAs are characterized as more site-specific and flexible to the local conditions. The following four cases can typically illustrate the diversity in the characteristics of Autonomous MPAs in Japan (Fig. 1.1).

7.3.1 Walleye Pollock MPA at Shiretoko

The first case is the Shiretoko World Natural Heritage Area. In this area, various fisheries management measures have been autonomously implemented by local fishers. These include seasonal no-take zones for Walleye pollock (*Theragra chalcogramma*) spawning stock, which have been implemented by local gillnet fishers since 1995. The no-take zones are reexamined every year in the light of the previous year's performance and scientific advice provided by the local research station. This MPA was thus the first type of Autonomous MPA, i.e., a resource management-oriented MPA.

However, after inclusion in the World Natural Heritage List in 2005, the boundaries of this MPA were extended with the aim of conserving the Steller Sea Lion

Fig. 7.1 Ukiyoe of a Yokohama coastal landscape by Utagawa Hiroshige (1797–1858)

population, which is an endangered species listed on the IUCN Red List and whose main food source in this area is Walleye pollock. This MPA is now officially incorporated into the Marine Management Plan of the Heritage Area as a part of the ecosystem-based management measures (Makino et al. 2009).

To sum up, this MPA was instigated by local fishers for Walleye pollock resource management, but is formally recognized as a conservation measure for an endangered species under the UNESCO World Natural Heritage framework. Another interesting feature of this case is the interlock between local and autonomous measures by fishers and official management measures by an international convention. Chapter 8 of this book gives a detailed description of its background and other marine ecosystem conservation strategies in this heritage area.

7.3.2 Seagrass Bed MPA Off Yokohama

Yokohama City is located on the western side of Tokyo Bay, which is the most urbanized bay in Japan (Fig. 1.1). Since the seventeenth century, Tokyo Bay has been famous as a production area of high quality fish for sushi, such as conger eel (Congridae), mantis shrimp (*Oratosquilla oratoria*), sea bass (*Lateolabrax japonicus*), smelt-whiting (Sillaginidae), dotted gizzard shad (*Konosirus punctatus*), short-necked clam (*Ruditapes philippinarum*), oval squid (*Sepioteuthis lessoniana*), etc. The Yokohama area was especially famous for conger eel and mantis shrimp. Figure 7.1

shows a wood block print of the Yokohama coastal area in the early nineteenth century, in which you can see fishers at work. According to fisheries ground maps from the late nineteenth century, the majority of the coastal areas around Yokohama were tidal lands and shallow bottoms covered by seagrasses.

After WWII, Japan experienced very rapid economic growth. Since the 1970s, seagrass beds, whose existence is crucial for the eggs and juvenile stages of fish and shellfish, have almost entirely disappeared around urban areas such as Yokohama. Today, nearly all the 140 km of Yokohama coastline has been developed for industrial or residential use. Only about half a kilometer of natural coastline is left.

In 1981, a volunteer group of scuba divers organized a sea bottom cleanup activity in Yokohama City. This group later became the main organizer of seagrass replanting activities in this area after a researcher at the prefectural fisheries research station found a local seagrass bed at the southernmost part of the Yokohama coast, and the group began to plant the seeds of the local seagrass in neighboring areas.

Today, seagrass recovery activities are enthusiastically conducted by a wide range of participants, from NGOs, local FCAs, local schools, private companies, central/local governments, and local research institutions (Citizen Group Meeting for the Promotion of Nature Restoration 2005). One of the notable features of this activity is the very wide variety of participants. Also, the areas where seagrasses have been reestablished are now protected as no-take zones under the Fishery Coordinating Committees Direction (Table 2.2), and a part of their activities are financially supported by the local and central governments. Their activities have successfully expanded the seagrass-covered areas. As evidence of the success of their activities, spawning by oval squid was observed in 2004 for the first time in 30 years.

7.3.3 *Rocky Shore Calcification in Kagoshima Prefecture*

The third case of an Autonomous MPA is the restoration of the rocky shore calcification phenomenon in Iwamoto in Kagoshima Bay (Fig. 1.1). In Kagoshima Prefecture, about 80% of the original seagrass beds are now estimated to have calcified, and, worse, a population explosion of sea urchins in calcified areas is preventing the recovery of seagrass beds (Tanaka 2010). In 1978, the Iwamoto area seagrass bed covered 36 ha. It was the biggest seagrass bed in the Kagoshima Bay at that time. However, it had shrunk to 10 ha in 1996, and to virtually nothing in more recent years. Calcification was rapidly taking over. A rapid increase was also observed in the number of sea urchins, such as black longspine urchins (*Diadema setosum*), which have a low market value and are not utilized by fishers.

This prompted Kagoshima Prefecture and the local FCA to take action to restore the seagrass beds. However, the local fishers were mostly elderly and the number and the distribution of sea urchins were so large that they could not think of an effective strategy. However, a local fisheries high school (Sect. 1.4.2), which held a lecture

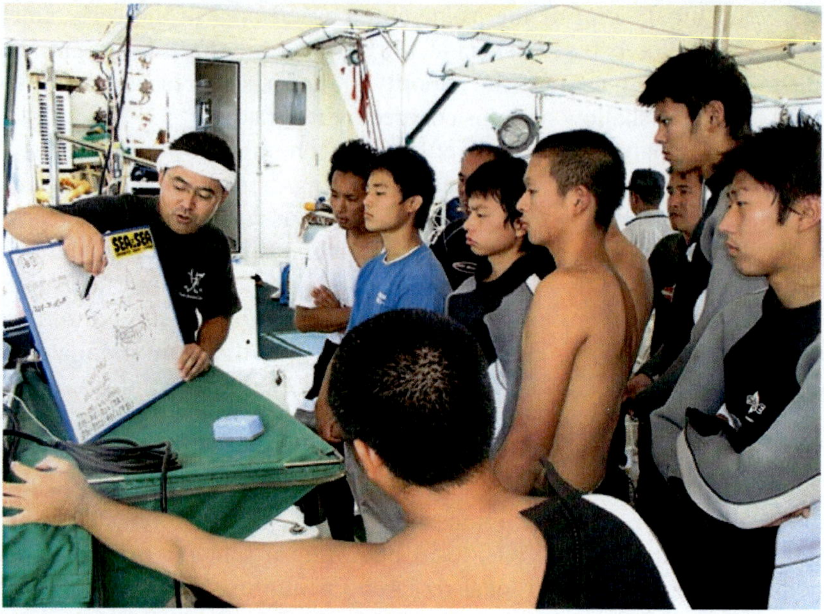

Fig. 7.2 High school students learning about the process of sea urchin culling

course on scuba diving, wanted to begin new activities that would give students a sense of social contribution. The local FCA, prefectural research station, and high school students have jointly planned a program to exterminate sea urchins, and have been planting seagrass in calcified areas since 2002 (Fisheries Agency 2007b).

The local government and scientists at the local research station coordinated these activities. Following procedures set up by scientists, students who have obtained diving licenses via their high school lecture course go underwater and remove the sea urchins (Fig. 7.2). The local fishers then plant seagrass embryos using mid-water nets. Regular monitoring and maintenance are then jointly conducted by high school students and local fishers. The seagrass meadows are reportedly spreading.

The distinguishing feature of this case is the educational purpose to the management activities. The collected sea urchins are also utilized for educational purposes at the high school. The goal of the activity is to encourage the recovery of seagrasses in this area and to observe the spawning of oval squid.

7.3.4 Coral Reefs in Okinawa Prefecture

The final example is from Okinawa Prefecture (Fig. 1.1). Okinawa Prefecture is famous for its tropical marine ecosystems, and is an important tourist destination. In the Zamami Islands, about 40 km west of Okinawa's main island, recreational

Fig. 7.3 A diver culling starfish on coral reefs in the Zamami area

diving and fisheries are key industries in the local economy. In the 1990s, popular diving spots in the Zamami area attracted hundreds of divers every day.

However, in the late 1990s, adverse effects on the coral reef ecosystems, caused by overuse, such as the anchoring of chartered boats, clumsy behavior by inexperienced divers, and sand disturbance, became apparent. To deal with this situation, a group comprising diving shops and the local FCA held discussions on coral reef conservation. In 1999, they set three strict no-take and no-diving MPAs (Kakuma 2007, 2009). Volunteer divers then regularly monitored the coral coverage rate based on a scientifically standardized reef checking method. They found that coverage increased from 30% to 50% in 3 years. A local research station supports their activities with scientific information and technical instructions.

However, in 2002, the population of the crown-of-thorns starfish (*Canthaster planci*) increased very rapidly within the MPAs. In one, it destroyed 70% of the coral reef. Intensive research and exclusion efforts are being conducted there, but no clear recovery has yet been reported (Fig. 7.3). Because these MPAs are no-take and no-diving zones, and since monitoring cannot be conducted very often due to its volunteer nature and cost limitations, the damage from the starfish was noticed only after several months, after considerable damage had occurred.

The feature of this case is that the local fisheries and tourism sectors are conducting conservation activities cooperatively. Also, one lesson is that, even in 100% no-take and no-entry areas, frequent monitoring is crucial.

7.3.5 Potentialities and Drawbacks of Autonomous MPAs

The above four cases are just small examples from many Autonomous MPAs all along the Japanese coast. Based on these local experiences, the Fisheries Agency compiled several guidelines for Autonomous MPAs set up on local initiatives (Fisheries Agency 2007a, b, c). The guidelines instruct how to organize and coordinate stakeholders, and reach consensus, as well as providing technical information.

The Autonomous MPA is an effective approach for several reasons: (1) local ecological and traditional knowledge, often conserved among local fishers, can be effectively applied with scientific support from local (prefectural) research stations; (2) the decision-making process is basically by consensus-building, which is very often on a unanimous basis, resulting in relatively low monitoring and enforcement costs, since mutual monitoring by local people/fishers is effective.

On the other hand, the Autonomous MPAs have the following drawbacks: (1) the emphasis on consensus-building can lead to cost-effective implementation, as explained above, but does not necessarily mean that the MPAs and other management measures implemented in the area will achieve the conservation objectives. The consensus-building approach often results in conservative rather than drastic and sweeping decisions. Therefore, continuous monitoring and adaptive revisions to the management plan based on scientific information are important. (2) Local fishers usually feel that autonomous rules made by themselves are as equally, or more strictly, binding as legal regulations. However, there is always a risk of moral hazard among participants, especially if a wide range of sectors is participating in the process. If all the participants were subject to moral hazard, Autonomous MPAs could be worse than top-down MPAs or Legal MPAs. Therefore, along with the decision-making and implementation authorities, the responsibility for conservation should be borne by the local organizations.

7.4 Social Aspects of Japanese MPAs

7.4.1 Objectives of MPAs

As the first principle of the Ecosystem Approach of the Convention of Biological Diversity points out, setting the objectives of marine ecosystem conservation is a highly social process (CBD 2000, see also Chap. 6). Different sectors of society view ecosystems in terms of their own economic, cultural, and social needs. Local communities and people are important stakeholders, and their rights and interests should thus be recognized. In other words, both cultural and biological diversity are central components of ecosystem conservation.

As explained in Chap. 1, the Japanese are fish-eaters, and fisheries products are their most important source of animal protein. The coastal areas near highly productive marine areas have been settled for thousands of years, during which many small-scale fishers have plied their trade. In these areas, local people have been living on marine ecosystem services, and human beings have become part of the local ecosystem.

As an example of this, to be discussed in Chap. 8, many components and most of the keystone species of the marine food web at the Shiretoko World Natural Heritage area have been utilized by local people for a long time. This means that, unless the objective of ecosystem conservation is to go back to the original wilderness thousands of years ago, responsible fisheries utilizing a wide range of species in a sustainable way in these areas are in fact working to conserve the ecosystem's structure and functions in this area. In other words, local fisheries are keystone species of the local marine ecosystems (Makino and Matsuda 2011). The seagrass bed MPA case in Yokohama described in the previous section also gives an insight into the sociohistorical background of use by the local people of ecosystem services. The coastal landscape in Fig. 7.1 depicts fishers and local communities living with coastal ecosystems several hundred years ago. Likewise, the goal of today's seagrass MPA activities is neither to abolish human exploitation, nor to return to the wilderness, but to achieve a sustainable balance among social and ecological systems.

Japanese citizens appear to support such a framework. For example, in January 2009, the Fisheries Research Agency of Japan conducted a social survey of 2,000 Japanese citizens, asking them to indicate the two most important human activities from a list of various marine uses (Fisheries Research Agency 2009). The result showed that 83% support fisheries production, followed by energy generation by tide or wind (54%), transport (21%), recreation (8.2%), etc.

This is why the governance processes of Japanese MPAs, particularly those of Autonomous MPAs, are closely linked to local fishers. In these cases, the objectives and nature of ecosystem conservation activities are different from the ones where people are immigrated to, so-called, new frontiers in relatively recent centuries. This shows the essential differences in social-ecological conditions between fish-eating countries such as Japan, Korea, and many other Asia-Pacific countries, and countries where the main food source is land-living species. Without understanding these important differences, it might be very difficult to hold constructive discussions on MPAs in the international arena.

7.4.2 Role of Local Fishers in Japanese MPAs

As explained above, local fishers are one of the most important participants in the process of setting up an MPA, even if it is the government that plans to set up Legal MPAs. However, stakeholders other than local fishers also participate in conservation activities, as exemplified by the Autonomous MPA cases described in the previous section. Actually, the participation of non-fishers in Autonomous MPAs that are initiated by local fishers is getting not rare in Japan. According to a survey report by the Fisheries Agency, more than 25% of conservation activities conducted by local FCAs are participated in by non-fishing people, such as environmental NGOs, local schools, local residents, private companies, etc. (Fisheries Agency 2007a). Local fishers are expected to play leading roles in setting, monitoring, and adaptively revising MPAs according to the local and ecological context.

However, as discussed in Chap. 6, it should be emphasized that ideal fisheries management is not always enough for the conservation of ecosystem structures and functions. The Autonomous MPAs initiated by local fishers are usually set up in the expectation of increased landing of valuable species, and tend not to pay due attention to species without market value. Certain gaps exist between fisheries management and ecosystem conservation. The nature of these gaps differs case by case and area by area, and their identification requires the participation of a wide range of stakeholders in ecosystem service use. This identification process is critically important, and once the gap is identified, it can be filled by environmental policy measures. Chapter 8 gives an example of an ecosystem conservation framework based on this approach and an estimate of its costs in the Shiretoko World Natural Heritage area.

In fact, it is an extremely difficult task to coordinate various sectors and interests. However, I think a clue can be found from the words of the leader of the environmental NGO participating in the seagrass bed MPA activities in Yokohama Bay. He pointed out that "even with very wide gaps in values or beliefs among participants, I think we can at least share the importance of education for local children… in that sense, we cannot omit children's experiences in catching and eating seafood in the field. This is very important." (Citizen Group Meeting for the Promotion of Nature Restoration 2005).

7.5 Other Activities for Ecosystem Conservation by Local Fishers

7.5.1 The Sato-umi Initiative

Conserving biodiversity means not only preserving wilderness, but also conserving the secondary natural environment with long-lasting human interactions. Based on this viewpoint, a new initiative called Sato-umi is now gaining increased attention. Sato- means community or village, and Umi means sea. Therefore, Sato-umi refers to seas that have been in long-standing interaction with human society. Its core vision is to realize societies in harmony with nature, i.e., built on a positive human–nature relationship.

The Sato-umi concept was first proposed by Dr. Tetsuo Yanagi (1998, 2007). His definition of Sato-umi is the coastal areas where human interaction has resulted in a high degree of productivity and biodiversity, and where a deep relationship between human life and traditional culture has led to the coexistence of humans and the natural world. Sato-umi comprises the elements of the coastal landscape that support fishery production and livelihoods: the seashore, tidal flats, seaweed beds, and fishing grounds.

The characteristics of this concept are a diverse mix of ecosystem types that produce a bundle of ecosystem services that vary according to specific social, economic, and ecological parameters. They are therefore context and/or place specific (McDonald 2009).

The concept of Sato-umi is supported in governmental policy frameworks. For example, the 3rd National Biodiversity Strategy of Japan (November 2007) called for appropriate conservation of Sato-umi. Also, the Basic Plan on Ocean Policy (March 2008) said,

> efforts should be made to embody the concept of Sato-umi, i.e., creating rich and beautiful marine zones and ensuring biodiversity while maintaining high biological productivity, by way of adding human labor at coastal marine zones in harmony with the natural ecosystem….In particular, with regard to coastal marine zones where biodiversity should be ensured while maintaining high biological productivity, the concept of Sato-umi should be emphasized from the perspective of preserving the marine environment.

Based on these political initiatives, the Sato-umi Creation Project was initiated by the Japanese Ministry of the Environment in July 2008. Currently, six pilot projects are being conducted with financial support from the government, including the Seagrass bed in Yokohama described in Sect. 7.3.

In addition, an international initiative on Sato-umi has just started. As the follow-up for the UN Millennium Ecosystem Assessment (2005), a multi-scale assessment with a number of components called sub-global assessments (SGAs) are now being conducted all over the world. SGAs are designed to meet the needs of decision makers at the scale at which they are undertaken, to strengthen global findings with on-the-ground reality, and to back up local findings with global perspectives, data, and models. In Japan, SGAs for Sato-yama and Sato-umi (Japan SGA) were launched in 2007 by the United Nations University's Institute of Advanced Studies to assess the ecosystem services derived from Sato-yama (village-mountain ecosystems) and Sato-umi and to provide a scientific base for actions to be taken toward conservation and sustainable use. The Japan SGA is the first integrated assessment in Japan where all dimensions, including social, cultural, economic, and ecological aspects, are incorporated and assessed. The main results of these assessments were announced at the 10th Conference of Parties of the Convention of Biological Diversity (CBD COP10) held in Nagoya, Japan, in October 2010.

7.5.2 The Uo-tsuki-rin (Fish Conservation Forest)

In Japan, it has been traditionally believed that "Without forests, there can be no fish." Local fishers have empirically known that good forests upstream create good fishing grounds downstream and where the river meets the sea. The Uo-tsuki-rin (fish conservation forest) is an old Japanese understanding that maintaining forests increases the productivity of fishing grounds. Today, these forests are known to prevent runoff from the land and provide clean, nutrient-rich water to the sea. There are many such forests around Japan. They have various local names such, as "fish-attracting forests," "fish-gathering forests," "fish-hiding forests," "fish mountains," etc. Records of fish conservation forests go back as far as the seventeenth century. Local fishers planted and managed upstream forests, and some feudal lords legally protected these forests and prohibited logging there (Umeno and Tanesaka 2001).

After the Meiji Revolution in 1868, the Former Forestry Law of 1897 was the first law to govern a Fish Conservation Forest. The current Article 25 of the Forestry Law of 1951 (as amended) inherited the prescription of the Fish Conservation Forests. Today, there are many locally initiated activities for Fish Conservation Forests. FCAs often participate in such activities as autonomous measures. As of March 2008, there were 58,000 ha of Fish Conservation Forests in Japan (Forestry Agency 2010). In 2005, the MAFF implemented a Support and Research Program for Environmental and Ecosystem Conservation Activities. Under this program, the national government financially and technically supports local fishers' projects for planting and maintaining fish conservation forests.

7.6 Discussion

The human dimensions or people-oriented factors such as social, economic, and institutional issues can dramatically affect the outcome of MPA implementation (Charles and Wilson 2009). The objectives of ecosystem conservation are a collective choice that varies from area to area and society to society. The fundamental concept of MPAs which suit fish-eating countries such as Japan appear to be local autonomous MPAs aiming for sustainable food provisioning as one of the most important ecosystem services. Fisheries rights and license systems that can effectively exclude outsiders comprise the key institutional framework for supporting the above system.

Concepts like Sato-umi and Uo-tsuki-rin could show a reasonable option for future ecosystem conservation activities worldwide. McDonald (2009) points out that interest in these concepts and recognition of their potential as a prototype for a sustainable system reflects the gradual shift in focus of the nature conservation movement: specifically, that from conserving designated protected zones and/or remote areas separate from human settlement to conservation wherein human intervention in nature is recognized as an integral element.

The scientific sector plays an important role in implementing MPAs. In many cases of Autonomous MPAs, the place, size, and timing of MPA setups are decided based on biological and technical support from prefectural research institutions. Science also plays an important role in monitoring. The local and traditional knowledge accumulated in coastal areas is of course an important source of ecosystem status recognition, but that kind of information is limited in space: in other words, to their fishing grounds. Official monitoring that covers wider areas and larger ecosystems is necessary for adaptive improvement of management strategies. The evidence of socioeconomic and ecological improvement by the introduction of

MPAs should be monitored and estimated based on sound science. Other scientific tasks for Japanese MPAs would be (1) to establish a theoretical and empirical foundation for adaptive management using MPAs as a tool for ecosystem conservation; and (2) to identify the differences in design/effects of MPAs in cold water ecosystems such as the Shiretoko World Natural Heritage Area (Chap. 8) and in tropical water ecosystems such as Kagosyma or Okinawa Prefecture.

References

Charles A, Wilson L (2009) Human dimensions of marine protected areas. ICES J Mar Sci 66(1):6–15

Citizen Group Meeting for the Promotion of Nature Restoration (2005) Nature restoration activities linking forests, sato (villages), rivers, and ocean: lessons from 13 advanced cases in Japan. Chuohoki Publishers, Tokyo (in Japanese)

Convention on Biological Diversity (2000) Decisions adopted by the conference of the parties to the convention on biological diversity at its fifth meeting, UNEP/CBD/COP/5/23

Convention on Biological Diversity (2004) Decisions adopted by the conference of the parties to the convention on biological diversity at its fifth meeting, UNEP/CBD/COP/7/5

Convention on Biological Diversity (2006) Decisions adopted by the conference of the parties to the convention on biological diversity at its fifth meeting, UNEP/CBD/COP/8/15

Dudley N (2008) Guidelines for applying protected area management categories. IUCN. http://data.iucn.org/dbtw-wpd/edocs/PAPS-016.pdf. Accessed 30 June 2010

Fisheries Agency (2007a) Guideline for actions against rocky shore calcification. National Association of Fisheries Infrastructures, Tokyo (in Japanese)

Fisheries Agency (2007b) Guideline for sea grass bed restoration activities. Marino Forum 21, Tokyo (in Japanese)

Fisheries Research Agency (2009) The grand design of fisheries and resource management in Japan. http://www.fra.affrc.go.jp/pressrelease/pr20/210331/. Accessed 30 June 2010

Forestry Agency (2010) Changes in the area of conservation forests. http://www.rinya.maff.go.jp/j/tisan/tisan/con_2_2_2.html. Accessed 30 June 2010

Kakuma S (2007) Diversity and multi-functions of Marine Protected Areas (MPAs) in Coral Reefs. Galaxea 8(2):91–108 (In Japanese)

Kakuma S (2009) Multi-functionality of MPAs in coral reef area. In: Yamao M, Shima H (eds) Multi-functionality of fisheries communities and fisheries in Japan. Hokuto-syobo, Tokyo, pp 89–110 (in Japanese)

Makino M (2010) Marine protected areas and local communities in Japan. Environ Res Q 157:55–62 (in Japanese)

Makino M, Matsuda H (2011) Fisheries diversity and ecosystem-based co-management, In: O. Rosemary, I. Pery, P. Curry, K, Cochrane (Eds.) Coping with global changes in social-ecological systems. Willey-Blackwells, London, pp 322–333

Makino M, Matsuda H, Sakurai Y (2009) Expanding fisheries co-management to ecosystem-based management. Mar Policy 33:207–214

McDonald A (2009) Concepts and practices of SATOYAMA SATO-UMI sub-global assessment in Japan. Presentation at East Asian Sea Congress, Manila, Nov 24

Millennium Ecosystem Assessment (2005) http://www.millenniumassessment.org/en/index.aspx

Tanaka T (2010) A review of the barren ground phenomenon and the recovery of seaweed communities in Kagoshima, southern Japan. Bull Kagoshima Prefect Fish Technol Dev Cent 1:13–18

Umeno S, Tanesaka E (2001) A history of the fish gathering forest dominated with the warm temperate evergreen broad-leaved tree plants; a case study in Kuroyama and Satori-Hirami sections on Kii-Oshima Island. Japan. Bull Kinki Univ Fac Agric 34:165–172 (in Japanese)

World Summit on Sustainable Development (2002) Plan of implementation of the world summit on sustainable development. http://www.un.org/esa/sustdev/documents/WSSD_POI_PD/English/WSSD_PlanImpl.pdf. Accessed 30 June 2010

Yagi N, Takagi AP, Takada Y, Kurokura H (2010) Marine protected areas in Japan: institutional background and management framework. Mar Policy 34:1300–1306

Yanagi T (1998) Coastal sea as a "Sato-umi". J Jpn Soc Water Environ 21(11):1 (in Japanese)

Yanagi T (2007) Sato-umi: a new concept for coastal sea management Ron. TERRAPUB, Tokyo

Chapter 8
The UNESCO World Natural Heritage List and Local Fisheries

Abstract People have been living on the *Shiretoko* peninsula for thousands of years. Even today, fishing is one of the main local industries in this area. Fisheries are an integral component of the local ecosystem, rather than unwanted extras to be eliminated from the "original ecosystem." The participation of the fisheries sector from the beginning of the process, scientific support from researchers, and accountability and official guarantees from administrators were the keys to constructing an effective system for ecosystem-based management in the UNESCO World Natural Heritage area. As a result, the total administrative costs for conservation measures were considerably decreased. Experiences from this case are potentially useful for informing ecosystem-based management in other countries where large numbers of small-scale fishers take a wide range of species under a fisheries co-management regime.

8.1 Introduction

The Shiretoko Peninsula, in northeastern Hokkaido, is the southernmost seasonal limit of sea ice in the Northern Hemisphere (Fig. 1.1). This region is characterized by closely linked terrestrial and marine ecosystems, and by a number of marine and terrestrial species, including several endangered species.

In 1964, the Shiretoko Peninsula and its surrounding marine areas were designated a National Park. In 1977, the Shiretoko 100 m^2 Campaign, which was a Japanese version of the National Trust, was launched to prevent land development and conserve native ecosystems; in 1994, local governments started a campaign to acquire UNESCO World Heritage nomination; and in January 2004, the national government formally nominated this area for the UNESCO World Heritage List. Shiretoko was added to the UNESCO World Heritage List in July 2005.

Since its nomination, various new measures have been implemented to conserve its exceptional ecosystems. The distinguishing feature of the approach taken here

M. Makino, *Fisheries Management in Japan: Its institutional features and case studies*, Fish & Fisheries Series 34, DOI 10.1007/978-94-007-1777-0_8,

was not to eliminate local fishers from the area, but to place their activities at the core of a management scheme designed to sustain ecosystem structure and function. Specifically, the existing fisheries management system was expanded to achieve ecosystem-based management. We call this the "Shiretoko Approach" (Makino et al. 2009).

Currently there are three UNESCO World Natural Heritage Sites in Japan: the Shiretoko Peninsula and its surrounding marine areas on Hokkaido Island (711 km^2), the Shiragami Mountains in northern Honshu Island (170 km^2), and Yakushima Island, south of Kyushu Island (107 km^2). The Shiretoko area is the largest and the only area that includes both land and sea. The Ogasawara Islands in the Pacific Ocean were nominated for the Heritage List in January 2010.

8.2 Shiretoko World Natural Heritage Area

8.2.1 Overview of the Ecosystems

The Shiretoko Peninsula and its adjacent marine areas (the Shiretoko WNH area) are considered to be the southernmost limit of seasonal ice floes in the Northern Hemisphere, and are affected by both the East Sakhalin cold current and the Soya warm current. The area has a complicated marine character created by these two currents, plus the intermediate cold water from the Sea of Okhotsk, and is home to a marine ecosystem in which a welter of organisms migrate and live (Ministry of the Environment and Hokkaido Prefectural Government 2007).

In early spring, the sea ice melts, and blooms of ice algae and other phytoplankton become the most characteristic part of the lowest trophic level of the Shiretoko ecosystems. The area's high productivity supports a wide range of species, including marine mammals, seabirds, and commercially important species (Sakurai 2007). In summer, sperm whales (*Physeter macrocephalus*) feed on squid in this area, and attract many tourists for whale-watching. In winter, numerous schools of Steller sea lion (*Eumetopias jubatus*) wander around the peninsula. Killer whales (*Orcinus orca*) also swim in the area throughout the year.

A distinguishing characteristic of this site is the interrelationship between its marine and terrestrial ecosystems. A mass of anadromous salmonids, such as chum salmon (*Oncorhynchus keta*), pink salmon (*O. gorbuscha*), masu salmon (*O. masou masou*), and dolly varden (*Salvelinus malma*), runs up the rivers in the peninsula to spawn. They serve as an important source of food for upstream terrestrial species such as the brown bear (*Ursus arctos*), Blakiston's fish-owl (*Ketupa blakistoni blakistoni*), Steller's sea eagle (*Haliaeetus pelagicus*), and white-tailed eagle (*H. albicilla*). The brown bear is the largest land animal in Japan, and occupies the top of the local ecosystem structure. The peninsula is also internationally important as a stopover point for migratory birds (IUCN 2005). Steller's sea eagles and white-tailed eagles migrate from Russia to this area in winter, but some white-tailed eagles live permanently on the peninsula.

This rich ecosystem and what it provides has also made it a human habitat for thousands of years. Stoneware unearthed from archaeological excavation sites has

Fig. 8.1 The tip of Shiretoko Peninsula (Photo by Makino)

numerous common features with those in the Northern part of the Eurasian Continent, suggesting that people immigrated here from Northern Europe via Siberia. Many clay pots and bones of Steller sea lions, seals, and fish have been found in the area. Between the eighth and twelfth centuries, other people advanced southward along the Okhotsk coast. Ainu culture developed here in the thirteenth to fourteenth centuries, and people lived mainly by hunting, fishing, gathering, and some agriculture. In the local Ainu language, the toponym "Shiretoko" literally means "earth [shir] + end [etok]," meaning the utmost end of the earth (Fig. 8.1).

In the Edo era (1603–1868), people from the Japanese main island began to commercialize fisheries products, etc., and increasingly controlled and suppressed the local people. After the Meiji revolution (1868), Hokkaido Island was formally incorporated into the Japanese territory. In 2010, a total of 19,059 people in 7,758 households were living in Shari and Rausu towns on the Shiretoko Peninsula. The main industries are fisheries, agriculture, and tourism.

8.2.2 Fisheries in the Shiretoko WNH Area

As described above, local people lived by fishing for many years. Commercial fisheries in Shiretoko began in 1790 with the foundation of a fishery market by the

Table 8.1 Catch composition in the Shiretoko Peninsula in 2008

Production volume		Production value	
Species	Tons	Species	JPY million
Salmon	34,076	Salmon	15,731
Walleye pollock	10,234	Walleye pollock	1,844
Atka mackerel	6,299	Atka mackerel	1,341
Squid	5,390	Cod	1,120
Cod	3,180	Squid	851
Flounder	1,198	Kelp	708
Others	3,326	Others	1,930
Total	63,703	Total	23,525

Source: Hokkaido Prefecture (2009)

rulers of mainland Japan. The main products at that time were dried or salt-cured salmon, trout, and herring (Shari Fisheries History Editing Committee 1979). After the Meiji revolution of 1868, offshore fisheries targeting halibut and cod started.

After WWII, the number of fishers in Shiretoko increased and a fisheries sector rapidly developed (Shiretoko Museum 2001). Today, the marine areas around the peninsula are among the most productive fisheries sites in Japan. In 2008, local fishers were yielding 63,703 ton of fish, worth 23,525 million yen. Table 8.1 shows the catch composition in Rausu town and Shari town in 2008 (Department of Fisheries and Forestry, Hokkaido Prefectural Government 2009). Their main target species and gear types are salmonids using set nets, common squid (*Todarodes pacificus*) by jigging, and walleye pollock (*Theragra chalcogramma*), Pacific cod (*Gadus macrocephalus*), and Atka mackerel (*Pleurogrammus azonus*) by gill netting. Fish processing industries are also very active here. The dried kelp (*Laminaria diabolica*) produced in this area is also highly prized, and fetches best prices on the Japanese market.

8.2.3 Steps to Achieving the UNESCO World Natural Heritage List

In January 2004, the government of Japan formulated a management plan and nominated the region for a UNESCO World Heritage Listing. The World Conservation Union (IUCN), the consultative body of UNESCO, then reviewed the proposal and management plan, and conducted a field evaluation in July 2004. After the field evaluation, the IUCN expressed the following concerns: (i) The level of protection of marine components was not high enough. In particular, walleye pollock, one of the main food sources of Steller sea lions, needed to be managed properly. (ii) The impact of construction along the rivers in the peninsula on the wild populations of salmonids, which link the marine and terrestrial ecosystems, required investigation. Then, in February 2005, the IUCN made two more suggestions: (iii) the marine

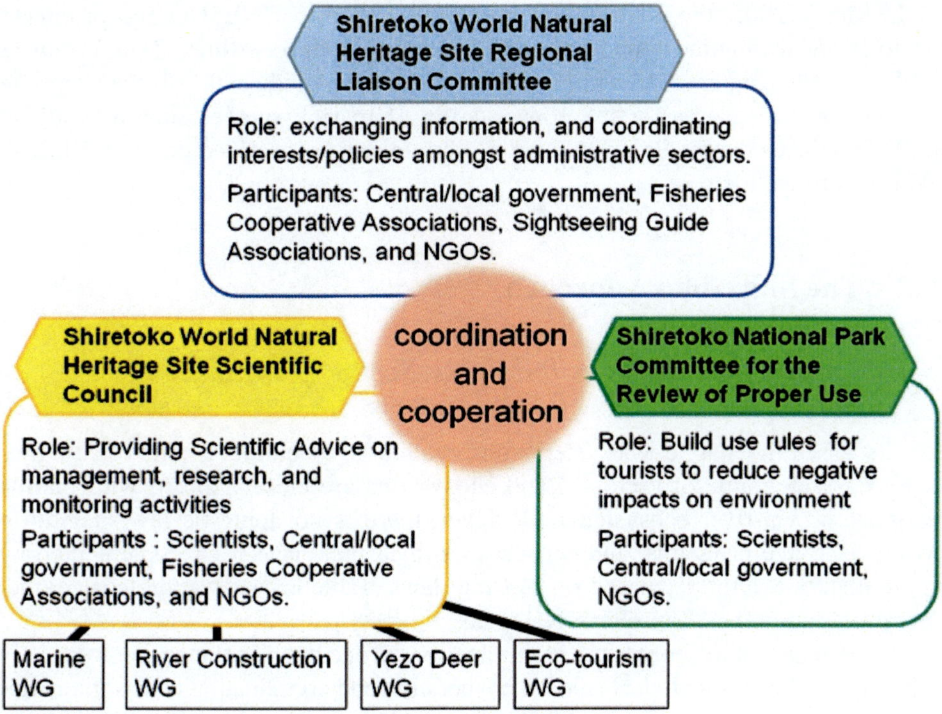

Fig. 8.2 Cross-sector coordinating system in the Shiretoko WNH

component of the site should be expanded, and (iv) a Marine Management Plan should be formulated as soon as possible to ensure the protection of marine species.

As shown in Table 8.1, walleye pollock is one of the most important resources for local fishing people in the Shiretoko area. Points (i) and (iv) from IUCN amplified worries on the part of the fisheries sector in the area. In addition, after the IUCN field evaluation report was made public, the mass media reported this issue in a sensationalist way. One TV station broadcasted a program entitled "Which is more important, fishers or sea lions?" Also, several scientists who did not know much about the actual situation recommended far stricter restrictions on local fisheries. Naturally, these reports played on the fishers' fears, and provoked some of them to strongly oppose the nomination.

Right after the arrival of additional suggestions from the IUCN in February 2005, core scientists in the Marine Working Group (Fig. 8.2) held discussions with leaders of local FCAs. Fishers thought it was a matter of life and death, and discussions inevitably became very heated. After repeated and intensive discussions, the Marine Working Group reached the firm conclusion that the Marine Management Plan, which was referred to in the IUCN suggestion point (iv), should set its objective as "to satisfy both conservation of the marine ecosystem and stable fisheries through the sustainable use of marine living resources in the marine area of the heritage site."

In March 2005, the government officially replied to UNESCO and promised (1) to extend the marine boundary from 1 to 3 km from the coastline, (2) to formulate a Marine Management Plan within 3 years, and (3) to include in the plan appropriate management measures for the conservation of marine species such as walleye pollock and sea mammals. In July 2005, Shiretoko was inscribed on the UNESCO World Heritage List.

8.3 The Shiretoko Approach

8.3.1 New Organizations for Cross-Sector Coordination

As discussed in Chap. 6, ecosystem conservation is, by nature, a suite of activities across a wide range of sectors. Land ecosystems are closely linked with marine ecosystems via river ecosystems. However, there is no domestic law specific to World Heritage programs, and conservation measures have been implemented by more than one authority based on separate laws. Table 8.2 shows the legal framework relating to the Shiretoko WNH area.

As in many other countries, administrative procedures in Japan are vertically structured. This often hinders smooth cooperation and coordination across ministries and departments. For example, the Natural Park Law of 1957 does not allow the Ministry of the Environment sufficient authority to regulate adverse impacts from fisheries activities on marine ecosystems. Because the fisheries sector has a long history as the mainstay of the regional economy, coordination with fisheries is especially important. The tourism sector, which is another important sector in the regional economy, experiences the same conditions. Therefore, a new system for cross-sector coordination was established for management of the Shiretoko WNH area (Fig. 8.2).

In October 2003, the Shiretoko World Natural Heritage Site Regional Liaison Committee was established with officers from a wide range of ministries and departments in central and local government. It discusses the proper management of the site, passes information back and forth, and coordinates various interests among sectors. Local Fisheries Cooperative Associations (FCAs), the tourism sector, the Scientific Council (described later), and NGOs are also members of this liaison committee. The committee serves as the core arena for policy coordination among administrative bodies.

The Shiretoko World Natural Heritage Site Scientific Council was established in July 2004. It provides scientific advice on the formulation of the management plan and on research and monitoring activities. The council had three working groups: the Marine Working Group for marine ecosystem management, the River Construction Working Group for improvement of river constructions, and the Yezo Deer Working Group for managing Yezo deer. The Scientific Council and working groups are composed of natural scientists, social scientists, and representatives of ministries and departments in central and local government, of FCAs, and of NGOs (the author belongs to the Marine Working Group).

Table 8.2 Major legal basis and administrative authorities for the Shiretoko WNH area management

Public services	Legal basis	Administrative authority
Fisheries management	Fisheries Law of 1949 Fisheries Resource Protection Law of 1951 Law Concerning the Conservation and Management of Marine Life Resources of 1996	Fisheries Agency (Ministry of Agriculture, Forestry and Fisheries)
Pollution control	Law Relating to the Prevention of Marine and Air Pollution from Ships and Maritime Disasters of 1970 Waste Management and Public Cleansing Law of 1970 Water Pollution Control Law of 1970	Coast Guard (Ministry of Land, Infrastructure, Transport and Tourism) Ministry of the Environment
Landscape conservation and material circulation	Law on the Administration and Management of National Forests of 1951 Natural Parks Law of 1957 Nature Conservation Law of 1972	Ministry of the Environment Forestry Agency (Ministry of Agriculture, Forestry and Fisheries)
Species protection	Law for the Protection of Cultural Properties of 1950 Law for Conservation of Endangered Species of Wild Fauna and Flora of 1992 Wildlife Protection and Appropriate Hunting Law of 2002	Ministry of the Environment Ministry of Education, Culture, Sports, Science and Technology

Every year about two million tourists visit the area. The Shiretoko National Park Committee for the Review of Proper Use, founded in 2001 and extended in 2004, has carried out research and discussions on proper-use rules for tourists. In April 2010, the Working Group for Eco-tourism was newly established under the Scientific Council.

Through these organizations and their interrelationships, stakeholder participation is ensured, information and opinions are exchanged, and consensus between the wide-ranging interests of multiple users of the ecosystem services is built, naturally increasing the legitimacy of the management plans and rules. This is the core institutional framework for integrated management under the Shiretoko Approach.

8.3.2 The Marine Management Plan

The Multiple Use Integrated Marine Management Plan (abbreviated to the Marine Management Plan) was developed by the Marine Working Group of the Scientific Council, and decided by the Ministry of the Environment of the Government of

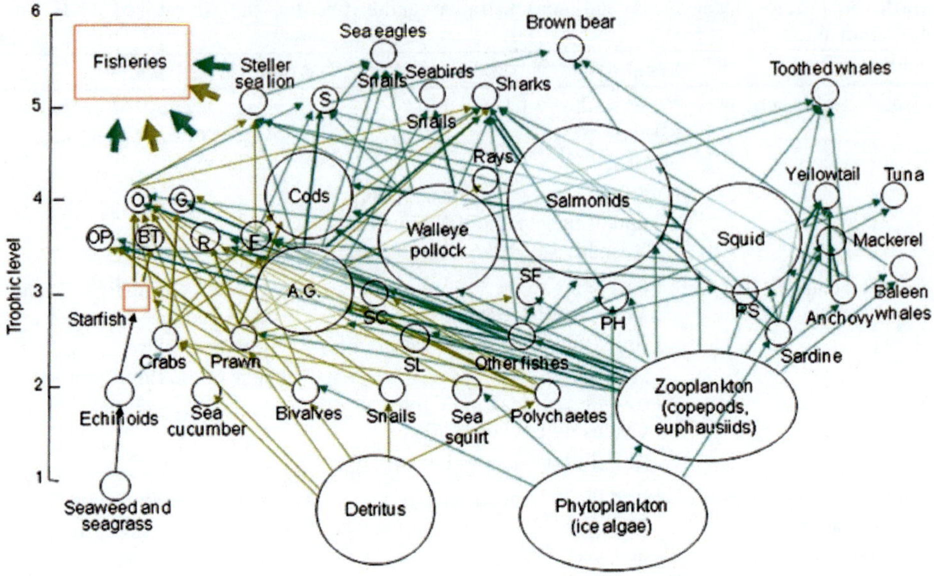

Fig. 8.3 Food web of the Shiretoko WNH area (as depicted by the Marine Working Group of the Scientific Council). *AG* arabesque greenling; *BT* bighand thornyhead; *F* flatfish; *G* greenling; *O* octopus; *OP* ocean perch; *PH* Pacific herring; *PS* Pacific saury; *R* rockfish; *S* seal; *SC* saffron cod; *SF* sandfish; *SL* sand-lance

Japan and the Hokkaido Prefectural Government in December 2007. The Marine Management Plan defines measures to conserve the marine ecosystem, strategies for maintaining major species, monitoring methods, and policies for marine recreational activities. The objective of the Marine Management Plan, which is stated at the beginning of the text, is "to satisfy both conservation of the marine ecosystem and stable fisheries through the sustainable use of marine living resources in the marine area of the heritage site." The fisheries sector has participated from the beginning of the drafting process. Because the ecosystem is disturbed, unclear, and complex, the Marine Management Plan stipulates the adoption of adaptive management (Walters 1986) as a basic strategy.

Monitoring is the key component in adaptive management. To monitor the Shiretoko marine ecosystem, the Marine Working Group drew up a food web (Fig. 8.3) and identified indicator species. The identified indicator species are salmonids (e.g., chum, pink, and masu salmon), walleye pollock, Pacific cod, Steller sea lion, etc. They are selected from keystone species, predators at higher trophic levels that are likely to have a great impact on ecosystems, and endangered species in the waters surrounding Shiretoko.

Under the Shiretoko Approach, the local fishers are acknowledged as an integral part of the ecosystem, and the data they provide are officially utilized to monitor the ecosystem cost effectively. Local FCAs have been collecting and compiling catch data for more than 60 years. These data cover many of the indicator species and other major marine species in the food web. For some species, such as walleye pollock, a great deal of detailed biological information, such as size, time and place of catch, and maturity, has been accumulated on an autonomous basis by pollock fishers. This information is an important baseline for monitoring changes in the functions and structure of the Shiretoko marine ecosystem.

Of course, catch data alone are not enough for monitoring the entire marine ecosystem, since fishers' behavior takes place in an economic context. Therefore, the Marine Management Plan specifies monitoring of noncommercial species, as well as basic environmental indices such as weather, water quality, sea ice, and plankton. Monitoring activities for these indicators are conducted by governmental agencies. In other words, the Shiretoko Approach initiated a system of role-sharing between government and ecosystem service users for sustainable and economically efficient ecosystem monitoring.

The multidimensionality and complexity of natural ecosystems and the human impact implies that not all environmental variables can be monitored and assimilated, and that indicators have to be used to summarize information of interest. Indicators are likely to be more easily understandable by the wide range of ecosystem service users than, for example, model outputs, and could provide the transparency required to promote dialogue (Degnbol and Jarre 2004). Cost-effectiveness concerns are included in the concept as well (National Research Council 2000; Cairns et al. 1993). Of course, if appropriate indicators are monitored, they will announce if something is wrong, but do not necessarily tell what is wrong, or what management action should be taken to mitigate its effects (Rochet and Trenkel 2009). As outlined by the FAO (1999), indicators are not an end in themselves, but have to be incorporated into a broader management framework. In the Marine Management Plan in the Shiretoko WNH, adaptive management is adopted as the basic management framework for this purpose.

Usually, an adaptive management framework determines the criteria and feedback control rules for indicator species: for example, monitoring of indicator species and implementation of conservation actions to maintain each species above a threshold abundance or to regain its abundance above a numerical goal by a given date (Matsuda et al. 2009). However, the current Marine Management Plan does not set these thresholds or numerical goals. In other words, we do not have clear decision-making rules for utilizing indicators in an adaptive way. The next task is to develop reference points representing the overall status and long-term trends of the ecosystem to be adaptively referred to in the overall management scheme.

8.3.3 Walleye Pollock and Steller Sea Lion

After the field evaluation in July 2004, the IUCN expressed concerns about the conservation of Steller sea lions and suitable management of its prey, the walleye pollock. As shown in Table 8.1, walleye pollock is the most important resource for local gill net fishers in the Shiretoko area.

The operations for walleye pollock are managed by licenses from the prefectural governor, and fisheries coordinating regulations that were established based on the Fisheries Law of 1949 and the Fisheries Resources Protection Law of 1951. This stock is also managed by the national government under the Total Allowable Catch (TAC) system, based on the Law Concerning the Conservation and Management of Marine Life Resources of 1996. In addition to official management by TAC, various autonomous management measures have been implemented, as explained below.

Local fishers compile data on catch size, time, area, body size, maturity, etc. These data are provided to the prefectural research station for analysis. The results are returned to the walleye pollock fishers, and management measures are discussed. For example, the local fishers voluntarily enlarged the mesh size of their pollock gill nets from 91 to 95 mm in 1990s, in accordance with research results provided by the research station.

Gill net fishers divide the fishery ground into 34 areas based on their local knowledge and experience. Since 1995, they declared seven of these areas protected to conserve resources. These protected areas include a portion of the walleye pollock spawning ground. The protected areas are reexamined every year on the basis of the previous year's performance and scientific advice from the local research station. After nomination for World Heritage Listing, another six areas were designated as protected (see Sect. 7.3.1).

Another example of an autonomous measure to conserve resources is the reduction of fishing capacity. The number of gill net vessels in the late 1980s was 193. To reduce fishing capacity in accordance with stock status, local fishers have decommissioned more than half of their vessels since 1996. Compensation for this decommissioning, about 1.1 billion yen, was jointly funded by the remaining fishers and the FCAs. The government bore the interest costs for borrowing compensation money from the bank. In 2002, fishers introduced a joint operation system to reduce fishing pressure by 20% and further reduce operation costs: five boats form a group, with each group suspending operations in turn.

Various autonomous measures are implemented for other resources as well. The local FCAs fund their own monitoring and research. Although these management measures are not well defined or described, they regulate the impact of fishing on stock. The Marine Management Plan officially incorporates these autonomous measures. An important next step is scientific verification of the validity of these measures.

Shiretoko fishers mainly catch the Nemuro stock of walleye pollock. In this respect, it is important to note that the Nemuro stock is shared by Russian trawlers operating around the southern Kuril Islands, where Japan and Russia have had a territorial dispute since the end of the Second World War. The total annual catch in the Shiretoko WNH area was around 100,000 ton in the late 1980s, but it has dropped drastically since 1990: in 2008, it was 10,234 ton. It seems likely that both environmental changes and increased fishing efforts in the late 1980s caused the stock collapse in early 1990s (Ishida et al. 2006). However, because of a lack of catch or biological data, the principal factor in the stock decline has not yet been identified by Japanese researchers.

Next, the Steller sea lions. The Okhotsk and Kuril population of the Steller sea lions migrate from their breeding and landing grounds in Russian waters to the Shiretoko WNH area for overwintering and foraging. Because the Asian population of Steller sea lions sharply declined until the 1980s, this species is classified as endangered on the IUCN Red List. Fortunately, its population has been gradually increasing, at 1.2% per year, since the early 1990s (Burkanov and Loughlin 2005). The entire population, which extends throughout the Sea of Okhotsk, the western part of the Bering Sea, and the Komandorskie Islands, was estimated at 15,676 in 2005 by enumerating the reproductive colonies. Based on this trend, Ministry of the Environment, Japan, ranked the sea lion as "vulnerable," the third rank of threatened species.

On the other hand, from the viewpoint of fishers operating in the Shiretoko WNH area, the Steller sea lion is a destructive animal. Sea lions swim inside the nets to eat the fish and then break the nets to escape. This damage continues to increase, and estimated damage to fishing nets now costs more than JPY 800 million in Hokkaido (Fisheries Agency and Fisheries Research Agency 2007). Therefore, to mitigate the damage, 116 Steller sea lions were culled each year under the Fisheries Law. However, since this cull size had no strong scientific foundation, in 2007, the Fisheries Agency of Japan revised the procedure for setting the cull limit. It is now calculated based on potential biological removal (PBR) theory (Wade 1998), which is used under the U.S. Marine Mammal Protection Act. Using the number of sea lions migrating to Japanese waters and the life history parameters used for the eastern Aleutian population, the calculated PBR was 227. Sea lions are also often by-caught in bottom set nets, gill nets, and set-net fisheries in Hokkaido, but there are no official statistics on the number of bycatches by these fisheries. The estimated number of total bycatches was between 55 and 107, and the highest number was subtracted from the PBR. Finally, the revised cull limit was set at 120 individuals. If information on bycatch sea lions is improved, the margin of error can be narrowed and the cull limit may be increased. Note that the culled sea lions are not wasted, but are consumed locally as food. In other words, they are marine resources for Japanese.

8.3.4 Interrelationship Between Marine and Terrestrial Ecosystems

Many anadromous salmonids, including hatchery-derived chum and pink salmon, return to rivers in Shiretoko to spawn. Upstream, they serve as an important source of food for terrestrial mammals and birds of prey, and contribute to biodiversity and nutrient circulation. Also, as shown in Table 8.1, salmonids are the most important fisheries resource for set-net fishers operating on the Okhotsk Sea. Under the Fisheries Law and Fisheries Resource Protection Law, set-nets are allowed only for fishers with Set-net fishery rights in marine areas, and other catching is prohibited in all rivers and near the mouths of certain rivers.

Man-made constructions such as dams risk obstructing wild salmonids in their escapement and preventing natural spawning. To maintain and facilitate interactions between marine and terrestrial ecosystems, man-made constructions along rivers in the Shiretoko WNH area have been modified since 2005.

The River Construction Working Group under the Scientific Council (Fig. 8.2) conducted a thorough survey on all the rivers in the area, and identified 118 man-made constructions in the Shiretoko WNH area. The working group then evaluated their impact on salmonids and investigated possible structural modifications, taking into account their effects on disaster risk for local residents. In some cases, modifications could lead to an increased risk of disasters in densely populated areas, and so they were not modified. As a result, 31 structures have been modified or are under modification as at the end of May 2010. To scientifically verify the effects of these measures, a 3 year program is in progress to monitor the upstream run, number and distribution of spawning beds, substrate composition, current velocity, and discharge. Preliminary results of the program show an increase in the number, size, and distribution of spawning beds.

8.3.5 Marine Recreational Activities

The Shiretoko Peninsula is a popular tourist destination in Japan, and tourism is an important contributor to the regional economy. Since its addition to the World Heritage List, the number of tourists has increased considerably. Tourists use marine areas for sightseeing, sea kayaking, private boating, scuba diving, and recreational fishing, among other uses.

However, there have been growing concerns that unregulated recreational use of marine areas may have adverse effects on their ecosystems. For example, passage by boats and unregulated feeding and watching at close range may affect the survival of seabirds and marine mammals. Many local fishers also complain of obstruction by tourists.

To prevent these negative impacts on the marine ecosystem and local fisheries, the Marine Management Plan prescribes that recreational activities are to be managed under rules formulated by the Shiretoko National Park Committee for the Review of Proper Use (Fig. 8.2). This committee is composed of academics, tourism and guide representatives, environmental NGOs, and officers representing forestry, the coastguard, the environment, and local government. The committee prescribes patrols and activities to monitor tourist use, formulates rules for tourists, and promotes ecotourism. In addition, a new working group for sustainable tourism, the Working Group for Ecotourism, was formed under the Scientific Council in April 2010.

8.4 Administrative Costs of Conservation Measures

As pointed out by Barnes and McFadden (2008), lack of financial resources is one of the most challenging barriers to implementing ecosystem-based management. Table 8.3 shows the estimated administrative costs for the ecosystem conservation

Table 8.3 Administrative costs for the Shiretoko WNH area in 2006

Cost item	Amount (JPY million)	Purpose
Running costs for Scientific Council and Working Groups	17.5	Giving scientific advice on management plan
Running costs for the Committee for the Review of Proper Use and Shiretoko Eco-tourism Association	15.1	Development of strategies for suitable tourism
Research and monitoring activities	54.7	Monitoring and research into adaptive management
River improvement	284.9	Modification of river constructions
Personnel	101. 8	Administrative staff at the Ministry of the Environment and Hokkaido Prefecture
Total	473.5	

Source: Makino et al. (2009)

measures in the Shiretoko WNH area in 2006, based on information provided by the Ministry of the Environment, the Forestry Agency, and Hokkaido Prefecture. These costs are additional expenses that result from addition to the Heritage List: they do not include conventional fisheries management costs. Also, the personnel accounted for in this table are engaged mainly in Shiretoko WNH affairs: five full-time and four part-time workers at the Ministry of the Environment, one full-time at the Forestry Agency, and seven full-time for Hokkaido Prefecture. The average wage of government officers was used to calculate these personnel costs.

In 2006, fisheries production was 22,966 million yen, and tourists spent an estimated 36,617 million yen in the area. The total administrative cost thus corresponds to 0.8% of the sum gained by the two principal industries that depend on services from the marine ecosystems. The total cost of effecting ecosystem-based management seems small in comparison.

8.5 Discussion

8.5.1 Assessment of the Shiretoko Approach

As explained in Sect. 8.1, the Shiretoko Approach is not to banish local fishers from the area. On the contrary: local fishers are an integral component of the ecosystem, rather than parasites of the "original ecosystem." Moreover, local fishers are not an influence to be managed or controlled, but are expected to play an indispensable part in ecosystem-based management. Put briefly, the existing fisheries management system was expanded to achieve ecosystem-based management. Also, in Chap. 6, I referred to the Ecosystem Approach of the Convention on Biological

Diversity, and discussed the institutional advantages of the Japanese fisheries management framework and necessary supplemental measures to expand it to ecosystem-based management. In this subsection, I would like to assess the measures taken in the Shiretoko WNH based on the results recorded in Chap. 6.

First of all, the fisheries management carried out in the Shiretoko WNH area is one of the most highly developed in Japan, and most of the institutional advantages listed in Table 6.3 are realized to a major extent. For example, the local fishers have adopted wide-ranging autonomous measures in addition to formal management, as exemplified by their management of the walleye pollock. Similar measures have also been adopted for other species such as salmon, Atka mackerel, cod, squid, kelp, sea urchin, etc. Planning and enforcement of such activities are conducted by local fishers based on their local ecological knowledge (Gadgil et al. 2003), but researchers in local and central governments and universities also provide support as scientific information. In the Shiretoko Approach, these decentralized activities are recognized and formally incorporated into the Marine Management Plan. The next task is to scientifically verify and support the autonomous management measures implemented by fishers. One major challenge is the establishment of "multi-scale and interlinked coordinating organizations" that are fine-tuned to the marine ecosystems in the Shiretoko WNH area. This challenge is closely related to the territorial dispute with Russia, and will be discussed later.

The following activities can be summarized as supplementary measures to expand existing fisheries management into ecosystem-based management (Table 6.3). Ecosystem perspectives are incorporated through the identification and monitoring of indicator species. Interactions between terrestrial and marine ecosystems are secured by modifying man-made constructions on the river. Science-based procedures to set a limit for the culling of the Steller sea lion have alleviated damage to fisheries without increasing the risk of extinction. The next task is, as mentioned earlier, to develop reference points representing the overall status and long-term trends of the ecosystem, to be adaptively referred to in the overall management scheme. Progress should be facilitated in the scientific understanding of interrelationships between fisheries operations, indicator species, and ecosystem structure, function, and processes.

Under the Japanese fisheries co-management system, coordination and stakeholder participation are limited to the fisheries sector only: no other marine ecosystem users are included in the decision-making process. Also, the autonomous rules implemented by local fishers are usually shared only within the fisheries sector, which often causes problems with the use of resources or areas. In the Shiretoko WHN case, a new coordination system was established, and a wide range of stakeholders from various sectors are now part of the decision-making process. This new system encourages the exchange of information and opinions, and thus strengthens the legitimacy of the management plans and rules.

It is worth noting that in the Shiretoko WNH case, governments saved considerable costs on management, especially for ecosystem monitoring. A new data collection and monitoring system was established by role-sharing between the government and ecosystem service users. Ecologically important areas such as the spawning grounds of walleye pollock are spatially managed and protected by local fishing people. Overall, it can be said that the current framework in the Shiretoko WNH area has so far achieved good ecosystem-based management.

8.5.2 Future Challenges

In the Shiretoko WNH area, territorial disputes with Russia have encouraged participation by the fisheries sector. Russian trawlers are much bigger (700–4,000 gross tons) than Japanese gill net vessels (10–19 gross tons), and they reportedly catch smaller individuals of walleye pollock (Fisheries Agency and Fisheries Research Agency 2007). As yet, there is no coordination between Japan and Russia to deal with this conflict. Shiretoko fishers hope that the World Heritage Listing will attract international attention to this situation and lead to some form of more effective management of walleye pollock in the near future. This is an important task for the national government. Resolving this cross-scale linkage of management (Ostrom et al. 2002) is important at the ecosystem level. Because ecosystems are inherently open, the Shiretoko ecosystems are closely linked with adjacent areas, so ecosystem management measures should be coordinated internationally where needed. Although there are serious territorial disputes over the territories, dialog between scientific groups can be the first step to a resolution (Crosby 2007).

Based on the above consensus between researchers and the national government ministries, a meeting between the Russian President and the Japanese Prime Minister was held in parallel with the 34th G8 Summit in Hokkaido. At this meeting, the Cooperative Program was signed with respect to Japan–Russia cooperation in the fields of conservation and sustainable use of neighboring areas to Japan and Russia, such as the Sea of Okhotsk. The content of this program includes the conservation and rational use of the marine and onshore sections of neighboring areas, the use of information regarding the ecosystems, evaluations of marine environments, surveys of the effects of climate change on the ecosystems, and the expansion of exchanges between relevant institutions and experts in Japan and Russia. Also, the "Amur Okhotsk Consortium" was established among Chinese, Japanese, and Russian researchers in 2009 (http://www.chikyu.ac.jp/AMOC/history.html) as a first step toward cooperation in environmental conservation of the Sea of Okhotsk and the Amur River Basin.

Another looming problem facing ecosystem conservation in the Shiretoko WNH area is climate change. As explained in Sect. 8.2, the Outstanding Value of Shiretoko is strongly related to the presence of seasonal sea ice. This influences the productivity of the marine ecosystem, which in turn influences the productivity and diversity of the terrestrial ecosystem. The effects of long-term climate change could have a significant impact on these ecosystems. Local fishers, researchers, and residents agree that the amount and the thickness of the sea ice have been rapidly decreasing in recent years.

The Report on the Reactive Monitoring Mission by the UNESCO and IUCN held in February 2008 pointed out the need to develop a Climate Change Strategy that includes the following activities: (a) development of a monitoring program which identifies both long- and short-term impacts of climate change and specifically monitors parameters such as the extent of sea ice and the impacts on populations of key indicator species; and (b) adaptive management strategies that could be applied to minimize any impacts of climate change on the value of the site (UNESCO and IUCN 2008). The Scientific Committee of the Shiretoko WNH is now preparing a strategy in response to this report.

These new challenges are, by nature, cross-sector challenges, and cannot be tackled effectively by isolated measures by separate ministries and agencies. In this regard, several initiatives to coordinate and integrate a wide range of measures have been formulated by the Japanese government. For example, the Japanese government has put together the National Biodiversity Strategy of Japan, in accordance with the Convention on Biological Diversity, which provides targets and directions for measures to ensure conservation and the sustainable use of biological diversity. In the Third National Biological Diversity Strategy and Action Plan meeting (2008), the Shiretoko WNH was cited as a successful case of marine biodiversity conservation (Government of Japan 2008). In addition, the Ministry of the Environment is now formulating their first strategy specifically for the conservation of marine biodiversity. As for marine policy coordination, which includes resource conservation, biodiversity conservation, and international coordination, the Basic Plan of the Integrated Ocean Policy was formulated in March 2008 (see Sect. 2.4.3). It is expected that these strategies and the basic plan will facilitate the formulation of integrated measures for the future challenges described above.

8.5.3 Lessons Learned

Several lessons on building consensus with the fisheries sector on ecosystem-based management can be learned. Initially, few local fishers welcomed the nomination for the World Heritage List, afraid that inclusion in the List would lead to additional regulations for the sole purpose of environmental protection. Therefore, before nomination for the World Heritage List in January 2004, the Ministry of the Environment and Hokkaido Prefecture promised local fishers that both conservation of the ecosystem and stable fisheries would be essential. This promise was also stipulated as an objective of the Marine Management Plan. The fisheries sector has participated from the beginning in all the coordinating organizations shown in Fig. 8.2 and in the drafting process of the Marine Management Plan. In addition, explanatory meetings have been held several times at local fishing communities. The participation of the fisheries sector and accountability and guarantees from administrators were the keys to building a consensus.

At the explanatory meetings held at local fishing communities, several core researchers of the Scientific Council, including the Chair of the Marine Working

Group, participated and gave full and open explanations to local fishers. Therefore, in the Shiretoko Approach, researchers are not only providers of scientific information and advice, but also facilitators of the use of scientific information on ecosystem conservation by local fishers. This role is similar to that of the extension officers described in Sect. 1.5. According to the national government official who was in charge of the Shiretoko WNH program at that period, these attitudes by leading researchers supported and encouraged cross-sector coordination and communication among officials from related ministries and departments.

The Shiretoko Approach is based on the Japanese fisheries management framework. At the beginning of the listing process, IUCN people said the fisheries management system was too complicated. I personally think the Japanese management framework might have been interpreted by them as "setting the wolf to guard the sheep (a similar proverb in Japanese is "to set the cat to guard the dried bonito")." Therefore, as a member of the Marine WG of the Scientific Council, I made considerable efforts to explain the Japanese institutional framework and its relationship with ecosystem conservation in an internationally understandable manner. According to Copes and Charles (2004), Japanese fisheries management system can be categorized as a kind of "community-based co-management," which acknowledges local fishers as the primary participants in management, and that the involvement and support of the broader communities is essential. It is open to considering a wide range of human needs in the community, and therefore lends itself to the implementation of a balanced mix of biological, social, and economic objectives. This fisheries institutional background in Japan naturally leads to a different ecosystem-based management framework from, for example, that of Iceland or New Zealand, where market-based individual transferable quotas are the central policy tool. There is no unique path toward conserving marine ecosystems and sustaining livelihoods. What is required is a careful assessment of the existing institutional frameworks and the potential role of the fisheries sector in marine ecosystem management.

At the UNESCO/IUCN Report on the Reactive Monitoring Mission held in February 2008, the mission team applauded the Shiretoko Approach as "an excellent model for the management of natural World Heritage Sites elsewhere (UNESCO and IUCN 2008)." We hope that the knowledge gained in setting up in the Shiretoko WNH can contribute to future ecosystem-based management in other regions where large numbers of small-scale fishing people utilize a wide range of species under a fisheries co-management regime.

References

Barnes C, McFadden KW (2008) Marine ecosystem approaches to management: challenges and lessons in the United States. Mar Policy 32:387–392

Burkanov NV, Loughlin TR (2005) Distribution and abundance of Steller sea lions, *Eumetopias jubatus*, on the Asian coast, 1720s–2005. Mar Fish Rev 67:1–62

Cairns J, Mccormick PV, Niederlehner BR (1993) A proposed framework for developing indicators of ecosystem health. Hydrobiologia 263:1–44

Copes P, Charles A (2004) Socioeconomics of individual transferable quotas and community-based fishery management. Agric Resour Econ Rev 33:171–181

Crosby MP (2007) Improving international relations through marine science partnerships. In: Nordquist MH, Long R, Heidar TH, Moore JN (eds) Law, science & ocean management. Martinus Nijhoff Publishers, Leiden, pp 271–293

Degnbol P, Jarre A (2004) Review of indicators in fisheries management-a development perspective. Afr J Mar Sci 26:303–326

Department of Fisheries and Forestry, Hokkaido Prefectural Government (2009) Hokkaido Suisan Gensei (current situation of fisheries in Hokkaido). Hokaido Prefecture, Sapporo (in Japanese)

Fisheries Agency and Fisheries Research Agency (2007) Stock assessment for nemuro stock of Walleye pollock. Stock assessment report in fiscal year 2006

Food and Agriculture Organization of the United Nations (1999) Indicators for sustainable development of marine capture fisheries (FAO technical guidelines for responsible fisheries 8). Food and Agriculture Organization of the United Nations, Rome

Gadgil M, Olsson P, Berkes F, Folke C (2003) Exploring the role of local ecological knowledge in ecosystem management: three case studies. In: Berkes F, Colding J, Folke C (eds) Navigating social–ecological systems: building resilience for complexity and change. Cambridge University Press, Cambridge, pp 189–209

Government of Japan (2008) Third national biodiversity strategy. http://www.biodic.go.jp/convention/The%20Third%20NBS.pdf. Accessed 30 June 2010

Hokkaido Prefecture (2009) Status of fisheries in Hokkaido of 2008 (Suisan Gensei)

Ishida R, Torisawa M, Shida O (2006) Fisheries and fisheries resource in Shiretoko (continental shelf area). Kaiyo Mon 38:626–631 (in Japanese)

IUCN (2005) Technical evaluation report, Shiretoko (Japan) ID no. 1993

Makino M, Matsuda H, Sakurai Y (2009) Expanding fisheries co-management to ecosystem-based management: a case in the Shiretoko world natural heritage. Jpn Mar Policy 33:207–214

Matsuda M, Makino M, Sakurai Y (2009) Development of an adaptive marine ecosystem management and co-management plan at the Shiretoko world natural heritage site. Biol Conserv 142:1937–1942

Ministry of the Environment & Hokkaido Prefectural Government (2007) The multiple use integrated marine management plan

Museum S (2001) Fisheries in Shiretoko. Association for Shiretoko Museum, Shari (in Japanese)

National Research Council (2000) Ecological indicators for the nation. National Academy Press, Washington, DC

Ostrom E, Dietz T, Dolsak N, Stern PC, Stonich S, Weber EU (2002) The drama of the commons. National Academy Press, Washington, DC

Rochet MJ, Trenkel VM (2009) Why and how could indicators be used in an ecosystem approach to fisheries management? In: Beamish RJ, Rothschild BJ (eds) The future of fisheries science in North America. Springer Science, Dordrecht, pp 209–226

Sakurai Y (2007) An overview of the Oyashio ecosystem. Deep Sea Res II 54:2526–2542

Shari Fisheries History Editing Committee (1979) Shari fisheries history. Shari, Shari Town (in Japanese)

United Nations Educational, Scientific and Cultural Organization & International Union for Conservation of Nature (IUCN) (2008) Shiretoko natural world heritage site, Japan (the report of the reactive monitoring mission, 18–22 Feb 2008)

Wade PR (1998) Calculating limits to the allowable human-caused mortality of cetaceans and pinnipeds. Mar Mamm Sci 14:1–37

Walters CJ (1986) Adaptive management of renewable resources. McMillan, New York

Chapter 9
Comprehensive Management and Future Scenarios for Japanese Fisheries

Abstract A policy report published by the Fisheries Research Agency of Japan suggests there are five major aspects to fisheries management: (A) resource and environmental conservation, (B) food provision, (C) industrial and economic development, (D) local community development, and (E) the promotion of culture and science. Taking this multi-objective nature into account, the report then suggests a basic scheme for comprehensive fisheries management. It also shows three theoretical future scenarios for Japanese fisheries that explicitly reflect different sets of values: a global competition scenario, a national food security scenario, and an ecological mosaic scenario. At the end of this chapter, I discuss the national citizens' policy demands based on the results of a web-based questionnaire survey.

9.1 Introduction

In April 2008, the Fisheries Agency of the Ministry of Agriculture, Forestry and Fisheries (MAFF) requested the Fisheries Research Agency (FRA) to conduct a study on the management design and policy framework for the future Japanese fisheries sector. In response to this request, the FRA organized a study committee, which published a final report in March 2009 entitled "The Grand Design of Resource and Fisheries Management in Japan" (hereinafter called the FRA Grand Design Report). The FRA Grand Design Report analyzed the advantages and problems facing Japanese fisheries, identified urgent tasks to be dealt with, and presented three policy options that explicitly reflect differences in sense of value. It also made a relative evaluation of these three options and their applicability to citizens' policy demands, based on the results of the questionnaire survey (FRA 2009).

This chapter describes the social background of this report. I then explain the concept of comprehensive fisheries management, one of the main messages of

M. Makino, *Fisheries Management in Japan: Its institutional features and case studies*,
Fish & Fisheries Series 34, DOI 10.1007/978-94-007-1777-0_9,

the Grand Design Report. The latter part of this chapter discusses the three future scenarios and citizens' policy demands for the future of Japanese fisheries.

9.2 The Social Background of the FRA Grand Design Report

9.2.1 The Momentum for Fisheries Reform

In July 2007, a policy report on Japanese fisheries' structural reform was published by an influential think tank called the Japan Economic Research Institute. This institute is funded by leading Japanese companies, institutes within leading companies, major banks, etc. The title of the report was "Strategic and drastic reform of fisheries that conserve Japan's fish diet should be expedited." According to the report, the main objectives of this report were to prevent resource depletion, to enrich fishers and local communities, and to supply Japanese citizens with safe and reliable fishery products. The report concluded with these four recommendations to the national government (Japan Economic Research Institute 2007):

1. Ensure full conservation of the environment and fisheries resources based on scientific knowledge
2. Immediately implement structural reforms to allow the entry of new capital, technology, human resources, and marketing capability
3. Drastically reallocate the national budget to achieve such structural reforms
4. Construct an efficient distribution system designed to link production with end-consumption of fisheries products

The report also suggested dozens of specific policy measures to be implemented by the national government. Among these, several measures have sparked heated debate within the fisheries industry, academics, and the government, particularly about the suggestion in the report that to achieve structural reforms, "entry barriers" to fisheries should be abolished and entry made open to anyone. The report refers to fisheries management in New Zealand, Iceland, and Norway, and suggested the formal introduction of a system of Individual Quotas (IQs) or Individual Transferable Quotas (ITQs) to rationalize Japanese fisheries operations. As described in the case studies in this book, fishers have autonomously introduced IQs or similar catch quota systems for dozens of species, in spite of no explicitly defined ITQs being instituted in Japan. This report suggested the formal introduction of IQs or ITQs as government policy.

The so-called "entry barriers" under the current institutional framework are the following. The local Fisheries Cooperative Associations (FCAs) or members of FCAs have first choice when it comes to distributing fisheries rights under Article 14 of the Fisheries Law of 1949. Article 18 of the Law of Fisheries Cooperative Associations of 1948 also sets several criteria for being a member of an FCA: for example, that the FCA member must be an individual fisher who lives in the local community and engages in fishing at sea for at least 90–120 days per year. A company can also be a member of an FCA, but it must have fewer than 300 employees and a total fishing vessel tonnage of between 1,500 and 3,000 gross tons, and the base location of the company must be in the local community, etc. As explained in Chap. 2, this system of prioritizing small-scale fishers/companies results from the post-WWII fisheries reforms that were designed to improve the economic status of fishers actually engaged in fishing operations, not the large rights holders who had previously monopolized the coastal fisheries. The Japan Economic Research Institute's report noted that the social environment had drastically changed in the last 60 years and that the existing institutional framework was now making it difficult to develop the fisheries industry.

The National Federation of Fisheries Cooperatives (Zengyoren), which is a national federation of small-scale fishers and processors, spoke out strongly against the Japan Economic Research Institute's report, especially the suggestion to make entry to the coastal fisheries open to large-scale companies and the adoption of ITQs. In March 2008, Zengyoren published a reply in which it stressed the importance of cooperatives for local small-scale fishing operators and local economies. The points in the Zengyoren report were as follows:

1. The decline in the economic efficiency of the fisheries sector must be tackled, but not through the unrestricted entry of large companies from outside local communities.
2. Coordination and agreements among various fisheries on operations and type of equipment have up to now been jointly conducted by local FCAs and local governments. The adoption of ITQs would destroy these historically established and verified systems, and would change the fisheries sector into a shortsighted profit-pursuing industry.
3. The current fisheries institutional framework has previously allowed entry of outsider companies into fisheries, provided they do not clash with local interests; therefore, the real aim of the Japan Economic Research Institute's report is to take over the coastal fisheries, especially the profitable fisheries such as set-net fisheries or aquaculture, and to exclude local small-scale operators.

9.2.2 Linking to a Political Movement

The basic idea and the recommendations for fisheries reform from the Japan Economic Research Institute were accepted by the then cabinet. This allowed fisheries reform to be added to the agenda of the Council for the Promotion of Regulatory Reform of the Cabinet Office of Japan. In this way, the momentum of fisheries reform, which had been ignited by the Japan Economic Research Institute, became a political movement.

The Council for the Promotion of Regulatory Reform, which is composed of representatives from the private sector and academics, is a temporary council, with a lifespan of 3 years, tasked with encouraging structural reform of the Japanese economy. It was set up in January 2007, following its precedent councils and having the same purpose. The current and preceding councils' political standpoint is generally perceived to be fostering deregulation and a market-oriented approach, and promoting "small government." For example, as one milestone in its activities, the council proposed privatizing the national postal service sector, and the deregulation of temporary worker dispatch services for the manufacturing sector. Both were actually implemented by the then cabinet in 2007 and in 2004, respectively, but the pros and cons of these reforms are still under discussion by the current cabinet.

In December 2007, the Council recommended considering the introduction of ITQs, deregulation of the fishing rights system with respect to eligibility, priority setting, transferability, etc. In accordance with the Council's recommendations, the Fisheries Agency set up various programs and advisory panels to draw up answers for each recommendation. For example, they held intensive discussions on the applicability of IQs/ITQs, the potential for increasing the number of species managed under the TAC system, investigation of the actual situation of fishing rights distribution in coastal areas, etc. In April 2008, the Fisheries Agency asked the Fisheries Research Agency (FRA) to conduct a comprehensive study on a grand design and policy recommendation for the future Japanese fisheries sector. In response to this request, in April 2008 the FRA organized a study committee of 17 specialists, and published a final report in March 2009 (FRA 2009).

9.3 Comprehensive Fisheries Management: Objectives

One of the main messages posed by the FRA Grand Design Report was the concept of comprehensive fisheries management. It consisted of three components of comprehensiveness: objectives, evaluation criteria, and measures.

9.3.1 Five Aspects of Fisheries Management

The fisheries sector plays many roles in Japanese society. The FRA Grand Design Report classified the objectives of fisheries policy as 16 subcategories, which are then classified into five aspects: (A) the resource and environmental policy aspect, (B) the food policy aspect, (C) the industrial and economic policy aspect, (D) the local and community policy aspects, and (E) the cultural and science policy aspect (Fig. 9.1).

The FRA Grand Design Report suggested that comprehensive fisheries management should take all these five aspects into account when planning the management policy. By the same token, fisheries policy evaluation should be conducted referring to its contribution to these five aspects. The real political issue, therefore, is how to set relative weights or priorities among them, as discussed later (Sects. 9.6 and 9.7). The details of each aspect are discussed next.

Fig. 9.1 The five aspects of Japanese fisheries management (From the FRA Grand Design Report)

9.3.2 The Resource and Environmental Policy Aspects (A)

As described in detail throughout this book, fishers and their organizations play a core role in the Japanese fisheries management regime, in both resource and ecosystem conservation. They are expected to continue playing this role in maintaining fisheries resources at the appropriate level and adopting recovery measures for overfished resources. Management measures should be based on both scientific and traditional knowledge, and implemented through transparent decision-making processes (A-1: Conservation and recovery of fishery resources).

The conditions that today's fisheries sector needs to cope with include fluctuations in resources and the environment, such as multi-decade species alternation (Sect. 5.2) and global climate change. Adaptive fisheries operations that respond to such fluctuations should thus be conducted (Walters and Hilborn 1976). Activities for environmental conservation by fishers, as presented in Chaps. 7 and 8, or other measures such as energy saving or reduction of environmental impact should be promoted to conserve the structure and function of the marine ecosystems and the ecosystem services they provide (A-2: Harmony with the ecosystem and the environment).

Under an international management framework, Japanese fisheries should take the leadership in resources and ecosystem conservation in waters crossing national borders or public waters. When necessary, new international fisheries management organizations, such as regional organizations or grassroots organizations, should be established through international cooperation (A-3: Establishment of international management frameworks).

9.3.3 Food Policy Aspect (B)

Improvement of the self-sufficiency rate has always been one of the most important policy issues in Japan. The Basic Plan on Fisheries Policy of 2007 set the target self-sufficiency rate at 65% and the target fisheries production at 5.68 million tons in 2017 (B-1: Increase in production and improvement of the self-sufficiency rate).

Food quality is no less important than quantity. Japan has had tragic experiences of public health damage caused by pollution of seafood, known as Minamata disease. We must make our best efforts never to repeat this tragedy. Supplying safe and pollution-free products to Japanese citizens is thus a fundamental part of fisheries policy. Providing reliable and easy-to-understand information to consumers is also essential for improving the people's dietary life, health, and welfare (B-2: Security of food reliability and safety).

Rapid increases in global demand for seafood are likely to result in higher food prices. Also, as pointed out above, natural fluctuations in catch are an intrinsic nature of fisheries, but can cause instability in seafood prices. Fisheries policy should aim to stabilize both production price and quantity of supply to the Japanese citizens now and in the future (B-3: Security of supply stability).

9.3.4 Industrial and Economic Policy Aspect (C)

People's needs are always changing. Any outdated institutional barriers risk impeding fast and timely responses to change. Institutional frameworks for the fisheries sector should be adjusted continuously, to flexibly match the people's consumption needs (C-1: Institutional response to changes in consumer needs).

Since it is an industry, economic efficiency is very important for the fisheries sector. Stable employment is also an important objective of the fisheries policy. An appropriate competition and employment system that can withstand external shocks needs to be promoted (C-2: Realization of an efficient and stable operational environment).

As previously explained in a case study on sea cucumber fisheries in Aomori Prefecture (Sect. 4.2), some Japanese fisheries products have brand value in overseas markets. Fisheries policy should promote certified Japanese fisheries products and promote their differentiation in the international market (C-3: Promotion of internationally competitive products).

As a labor policy in the fisheries sector, a safer and cleaner working environment should be established that will attract new workers to join the fisheries sector, drawn by good working conditions (C-4: Improvement of the working environment).

9.3.5 Local and Community Policy Aspect (D)

Fisheries communities have been playing many roles other than food production. First of all, the fisheries sector creates diverse job opportunities in more than 5,000 fishing communities all along the Japanese coast. In fact, fisheries sectors, including processing and transporting, are the major source of income in many isolated coastal communities. In many of these communities, fishery-related jobs appropriate for each age bracket are available. Elderly people who have retired from their former occupations can come from other areas to join the workforce in fishing communities (D-1: Job creation for fishing community people).

In Japan, many coastal community infrastructures such as roads, ports, and wastewater treatment facilities have been developed by the Fisheries Agency, using the fisheries budget. To support rural living, good residential conditions and an attractive living environment should be included in fisheries-related land policy (D-2: Infrastructures and welfare development).

Local fishers have played a core role in many autonomous activities in fisheries management. Their autonomous activities are the key to reducing the total administrative costs of fisheries management. As exemplified in Chaps. 7 and 8, they can also help promote ecosystem conservation activities. They have also coordinated the use of the sea surface with other sectors, and supported rescue operations during disasters such as earthquakes or floods. National border surveillance is another important role of coastal fishers (D-3: Integrated coastal zone management).

9.3.6 Cultural and Science Policy Aspect (E)

Japan is a country with a long history of eating fish, and its people are blessed with a mosaic of cultures rooted in and unique to each local marine ecosystem. They include locally unique lifestyles, local knowledge, fishing techniques, fishery-related arts, local cuisine, etc. An important aspect in the social role of fisheries is that of enriching both rural and urban people's lives (E-1: Promotion of fisheries and fishing community culture).

To keep developing these cultural values, the features and attractions inherent in fisheries communities should be enhanced with opportunities for leisure, recreation, and education, and should prosper alongside other marine industries (E-2: Promotion of leisure and education).

Fisheries contribute to the collection of basic information for resources and ecosystems (see Chap. 8) and advancement of related sciences and technologies. They also make contributions to international society through diffusing such knowledge and technologies (E-3: Promotion of science and international contributions).

9.4 Comprehensive Fisheries Management: Evaluation Criteria

The second component for comprehensive management is the set of criteria used to evaluate fisheries management. Economic efficiency in monetary terms is the most often used criterion, and its importance cannot be overemphasized. However, we should realize that it is just one of several criteria for evaluating comprehensive management, and is applicable to a part of the objectives described in Fig. 9.1. The monetary value is a value for exchanges and cannot be completed itself in the policy discussion context. The analytical methodologies for economic efficiency evaluation are the most highly developed in the field of social science, and we should fully utilize their advantages. At the same time, we need to distinguish analytical performance as a scientific tool and its comprehensiveness for use in policy discussions.

There is no unique criterion for comprehensive fisheries management. It needs a suite of evaluation criteria. For example, Hillborn (2007) presents four criteria for various policy objectives: the biological criterion (maximum sustainable yield or MSY), the economic criterion (maximum economic yield or MEY), the social criterion (maximum job yield, or MJY, indicating generation of job opportunities for local communities), and the political criterion (minimum sustainable whingeing for reduction of political complaints, MSW). He concludes that the failure of the past MSY criteria is due to the success of other criteria or as a result of competition among criteria. However, these criteria are still designed for optimization purposes, meaning that maximization or minimization are set as the fundamental principles of evaluation, and then applied to various objectives. It is intuitively understandable to people with economic training, but there appear to be other equally important principles of evaluation in the policy discussions.

Following Miyagawa (1994), The FRA Grand Design Report suggested the following six criteria: efficiency (comparison between the results and the effort input or costs, such as economic efficiency, efficiency of employment, etc.); effectiveness (to what extent the objective has been achieved); sufficiency (to what extent the need is satisfied); fairness (distribution of costs and benefits); responsiveness (whether specific needs or values are satisfied); and appropriateness (appropriateness for society). Also, the resilience of the management system, as detailed in the next section, is another important criterion for fisheries management. Depending on the situation and problem recognition on the real field, a concrete and detailed set of criteria should be tailored on a case-by-case basis. Although the quantitative evaluation methods in some of the above criteria are still in their developing stages, it is important to try to ensure a comprehensive evaluation.

9.5 Comprehensive Fisheries Management: Measures

The final component of comprehensive fisheries management concerns the resilience and stability of the management. Charles (2007) pointed out that each management measure has its advantages and disadvantages, and that a combination of management measures (a "management portfolio" in Charles's term) that mutually reinforce the gaps and disadvantages reduces the risk of failure. Following this line, the FRA Grand Design Report accepts that a combination of management measures, rather than a single measure, is resilient to ecosystem changes, social changes, and uncertainty and diversity in fisheries.

The specific contents of a management portfolio must be determined based on the structure of a specific problem, the degree of urgency, the public demand for policies, and other factors in the real world. As a generalized scheme for deliberating the combination of measures, the FRA Grand Design Report proposes two types of typologies in fisheries management measures: the purpose-based typology and the approach-based typology.

9.5.1 Typologies for Fisheries Management Measures

Depending on what part of the fisheries system at which each measure is targeted, management measures can be divided into eight categories: (A) input control, (B) output control, (C) artificial seeds release, (D) conservation/restoration of ecosystems, (E) improvements in business structure, (F) improvements in postharvest treatment/processing/distribution, (G) development of human and organizational capacity, and (H) promotion of science and technology (Fig. 9.2).

On the other hand, the same management measure can be implemented in a variety of ways. Depending on the approach to implementation, measures can be divided into five categories: (1) the administrative approach, (2) the economic approach, (3) the information-based approach, (4) the judiciary approach, and (5) the autonomous approach. The nature of each management category, and its advantages and disadvantages, are described as follows.

Table 9.1 summarizes all the fisheries management measures grouped into the management categories described in this section.

A. Input control for resource conservation

   The first category is measures for input control, i.e., control of quality and quantity of fisheries operations. Typical examples in this category are restrictions on the number of vessels or type of equipment, engine power, days at sea, mesh size, etc. The advantages of these measures are summarized as follows. First, fundamental and long-term effects are expected. They can cover two or more species at the same time. The life cycle or growth stages of the target species can be clearly incorporated. Differences among fisheries sectors targeting

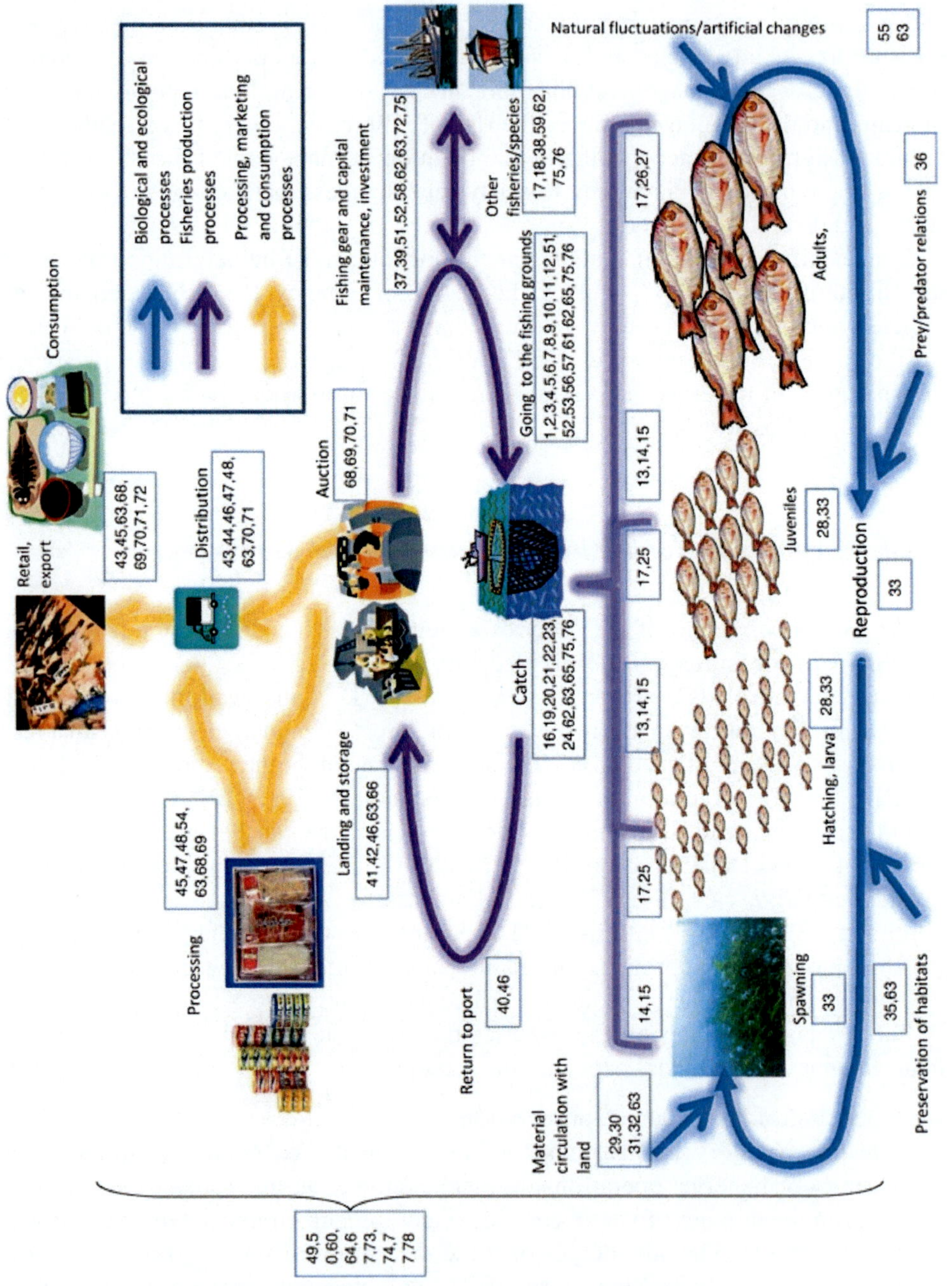

Fig. 9.2 Fisheries system and management measures. The numbers in the figure correspond to the measures in Table 9.1

Table 9.1 Summary of measures and categories

(a) Purpose-based classification				
A. Conservation of resources (input control)	Quantitative	Equipment		(1) Restriction on total tonnage of fishing boats; (2) restriction on horsepower of fishing boat engines; (3) restriction on the size of fishing gear; (4) restriction on fish tank capacity; and (5) restriction on light power
		Operation	Nontransferable	(6) Restriction of fishing effort (no. of fishing days, no. of fishing operations, no. of nets, etc.); (7) IEC (individual effort quota); (8) GEG (group-specific effort quota); (9) IOQ (individual oil quota)
			Transferable	(10) ITEQ (individual transferable effort quota) (with/without transfer restrictions, with/without deadline); (11) GTEQ (group-specific transferable effort quota) (with/without transfer restrictions, with/without deadline); (12) ITOQ (individual transferable oil quota) (with/without transfer restrictions, with/without deadline);
	Qualitative	Equipment		(13) Restrictions on fishing gear or fishing methods (restrictions on the type of fishing gear or fishing method, mesh size restriction, compulsory use of certain fishing equipment, etc.)
		Operation		(14) Restriction on fishing waters or timing (closed fishing area, closed fishing season, marine reserve); (15) rotational use of fishing grounds
B. Conservation of resources (output control)	Quantitative	Overall		(16) TAC (total allowable catch); (17) water- and season-specific TAC; (18) fishery- and fishing method-specific TAC
		Individual quota	Transferable	(19) IQ (individual quota); (20) IVQ (individual vessel quota); (21) GQ (group-specific quota)
			Nontransferable	(22) ITQ (individual transferable quota) (with/without transfer restrictions, with/without deadline); (23) ITVQ (individual transferable vessel quota) (with/without transfer restrictions, with/without deadline); (24) GTQ (group-specific transferable quota) (with/without transfer restrictions, with/without deadline)
	Qualitative			(25) Restrictions on size of catch (length); (26) restrictions on gender of catch; (27) restrictions on fishing of mature fish

(continued)

Table 9.1 (continued)

C. Addition and enhancement of resources		(28) Release of seeds
D. Conservation and restoration of ecosystem	On land	(29) fish conservation forest; (30) water quality management; (31) dam repair; (32) sediment and quicksand management
	In the water	(33) Conservation and restoration of seaweed beds and wetlands; (34) seabed tillage; (35) installation of fishing banks; (36) pest control or thinning
E. Improvement of business structure		(37) Promotion of reduction in fishing boats; (38) promotion of shifting to other fishery jobs or operation of two or more jobs; (39) reduction of capital by mini fleet arrangement
F. Improvement of postharvest treatment/disposal, processing, and distribution	Onboard	(40) Improvement of onboard treatment
	After landing	(41) Price sustenance and adjustment storage; (42) Development of fishing ports and markets; (43) promotion of exports; (44) rationalization of distribution chains; (45) improvement of added value by new product development, etc.; (46) quality standardization by hygiene standards, etc. (brand value enhancement); (47) reduction of distribution cost; (48) accumulation and improvement of processing and distribution technologies
G. Development of human and organizational capacity		(49) Establishment and change of management organization; (50) development, use and recruitment of human resources
H. Promotion of science and technology		(51) Development of fishing gear; (52) development of fishing methods; (53) development of fishing grounds and resources; (54) development of new processing methods; (55) understanding, evaluation and prediction of natural ecosystem mechanism

(b) Approach-specific classification		
1. Administrative approach	Legal protection	(56) Grant of fishery right
	Regulation/restriction	(57) Issuance of license; (58) setting of restrictions, regulations and procedures
	Guidance/order	(59) Adjustment among fisheries; (60) administrative guidance and diffusion; (61) stop order; (62) instructions from the committee or supportive order; (63) introduction of commodities and equipment contributing to reduction in environmental impacts
2. Economic approach	Promotion	(64) Dividend from subsidy, incentive pay, or membership fee
	Mitigation	(65) Collection of tax, levy, membership fee
	Neutral	(66) Pool system; (67) utilization of external private capital
3. Information approach	Promotion	(68) Branding; (69) eco-labeling
	Mitigation	(70) Black-listing; (71) approved list
	Neutral	(72) Business reports and press releases
4. Judiciary approach	Private	(73) Caveats, damage claims, etc.
	Public	(74) Criminal-code punishment and administrative punishment
5. Autonomous approach	Officially prescribed by the law	(75) Resources management agreement; (76) fishing grounds use agreement; (77) regulations based on the FCA's rules and FMO's rules
	Voluntary commitment	(78) Other voluntary regulations that go further than legal restrictions

the same species can be incorporated in a relatively easy way. They are also fairly robust to uncertainties in resources assessment. On the other hand, input control measures are inappropriate for flexible and fast management. Evaluating their effectiveness is difficult in many cases. Gaining a precise understanding of fishing pressure or angling pressure is sometimes difficult due to effort creep (Willen 1979), etc.

B. Output control for resource conservation
These measures control quality and quantity of catch, such as Total Allowable Catch (TAC), minimum size limit, etc. When scientific resource assessment is available, it is useful for flexible management depending on resource fluctuations. They are also relatively easy to understand, and readily applicable to management across borders or distribution over wide areas. If quotas are individually allocated as IQs, ITQs, or IVQs, the race for fish can be minimized and a coordinated operation can be anticipated.
However, the cost of setting up and implementing quota management is generally high. It is not possible to conduct fine-tuned management to match a species' life cycle. Also, when resources fluctuations or assessment errors are large, the success rate falls off rapidly. More important for coastal fisheries is that it is difficult to apply to multispecies fisheries. For individual quotas, initial distribution can be a big issue, and the costs of monitoring high grading or misreporting will increase.

C. Artificial seeds release for resource enhancement
As described in Sect. 3.4, artificial fish seed release, or fish ranching, is often adopted in Japan to help depleted resources to recover. These measures can directly improve locally decreased resources, and fishers easily understand and support these activities. Also, these measures often result in the improvement in fishers' consciousness for resource conservation. Therefore, it is good for the beginning of management processes. However, its long-term effectiveness is generally not clear. It is also not applicable to widely distributed species and is limited to species for which fish seed production technology has been developed. Finally, it is very difficult to assess changes in interspecies relationships, or potential impacts on the ecosystem such as changes in genetic diversity.

D. Conservation and restoration of marine ecosystems
These measures aim to improve the productivity of fishing grounds. Typical measures include replanting seagrass and controlling water quality. They can provide a wide range of functions through the conservation of marine ecosystems structures and functions, but the effects achieved and when they will occur are unclear.

E. Improvements to the business structure
Measures for reducing fishing capacity, promotion of alternative work, or catching other target species are categorized in this group. Net income will increase due to reduced costs or increased revenue. On the other hand, if the number of fishing boats is reduced, the financial burden of compensation is large. Also, when changing or enlarging the range of target species, it is generally difficult to coordinate with related fisheries sectors.

F. Improvements to postharvest treatment, processing, and distribution
The expected outcomes of these measures are improvements of fish prices, cost and added value, and more efficient use of resources. Fishery operators, processors, and distributors will enjoy increased income. They will also be able to cope better with varying consumer needs. But we do not know what direct effects will occur to the resource status or when such effects will occur.

G. Development of human and organizational capacity
Development and recruitment of management-related human resources or reform and reinforcement of Fisheries Management Organizations (FMOs) would allow them to cope flexibly with various problems in an innovative way. It also contributes to the formation of an appealing local community. However, human resources development inevitably takes time. Also, organizations tend to be conservative and exclusive if they become rigid.

H. Promotion of science and technology
Development of new technologies and new resources, and understanding and predicting ecosystem mechanisms can solve problems in a medium- and long-range and fundamental manner. However, rapid response to emergent issues is generally difficult. Even though relatively large initial investments are necessary, the feasibility of expected results is uncertain.

1. The administrative approach

   The government can directly implement various management measures. This approach is highly legitimate and stable because of its legal basis. Once implemented, it is effective as a fast-working measure to local problems. However, when subjects are dispersed, small in scale and diverse, monitoring is difficult, and in many cases, is more expensive than other methods. Government decisions also generally take time, making flexible and fast responses difficult under this approach.

2. The economic approach

   Economic incentives are one of the most powerful driving forces behind the behavior of fishers. Management measures based on economic incentives promote creativity and innovation, and work rapidly. Also, it is generally less costly than the administrative approach.

   On the other hand, it is not always known if it will have a direct effect on the target. Also, the market mechanism itself cannot deal with distribution problems, so supplementary measures might be necessary to ensure socially appropriate and fair distribution.

3. The information-based approach

   Disclosure or provision of information to a wide range of stakeholders, including fishers, local residents, and consumers, can be another approach. To fully utilize the measures in this approach, additional measures are needed to ensure the credibility of the information.

4. The judiciary approach

   This approach involves penalties or orders based on legal judgments by courts. It is highly legitimate and stable because of its legal basis, and effective as a

Table 9.2 Tool-box of fisheries management measures

					1. Administrative approach		
					Legal protection	Regulation/ restriction	Guidance/order
Conservation of resources	A. Input control	Quantitative	Equipment			1,2,3,4,5,58	59,63,60,62
			Operation	Nontransferable	56	57	59,60,62
				Transferable	56		
		Qualitative	Equipment			58,59	63,60,62
			Operation		56	14,57,59	60,61,62
	B. Output control	Quantitative	Overall			16,17,18	60,62
			Individual quota	Nontransferable	56		60,62
				Transferable	56		
		Qualitative				25,26,27	60.62
C. Addition and enhancement of resources							60
D. Conservation and restoration of ecosystem			On land			30,31,32	29,31,32, 63,60
			In the water			33	33,34, 60,63
E. Improvement of business structure							37,38,39,60, 62,63
F. Improvement of postharvest treatment/ disposal, processing, and distribution			Aboard the boat				40,60,63
			After landing				43,44,45,46,48, 60,63
G. Development of human and organizational capacity					49,50	49,50	49,50,60,62,63
H. Promotion of science and technology							51,52,53,54,55,60

deterrent. However, court judgments are generally time consuming and costly. The burden of proof very often becomes a big issue for expeditious and timely responses.

5. Autonomous approach

   Stakeholders can autonomously implement measures affecting issues of interest to them. This approach flexibly copes with individual specific cases, and requires lower administrative costs. However, in some cases, autonomous measures do not have strong binding and enforcing power.

9.5.1.1 Toolbox of Fisheries Management Measures

Table 9.2 shows the toolbox, or the matrix, of management measures, which plots target-based categories in the left-side column and approach-based categories in the top row to create specific examples of management measures.

When combinations of management measures in the real world are reviewed, it is crucial to make a full review of whether a combination of measures effectively

2. Economic approach			3. Information approach			4. Judiciary approach		5. Autonomous approach	
Promotion	Mitigation	Neutral	Promotion	Mitigation	Neutral	Private	Public	Officially prescribed	One-way commitment
64	65		71	70	72	73	74	1,2,3,4,5,58, 59,75,77	1,2,3,4,5,58,59,78
64	65	6,7,8,9	69,71	70	72	73	74	6,7,8,9,59,75, 76,77	6,7,8,9,59,75,78
	65	10,11,12	69,71	70	72	73	74		
64	65		68,69,71	70	72	73	74	13,58,59,75,77	13,58,59,78
64	65		68,69,71	70	72	73	74	14,15,59,75, 76,77	14,15,59,78
64	65	66	68,69,71	70	72	73	74	75,76,77	78
64	65	19,20,21	68,69,71	70	72	73	74	19,20,21,75,77	19,20,21,78
	65	22,23,24	68,69,71	70	72	73	74		
64	65		68,69,71	70	72	73	74	25,26,27,75,77	25,26,27,78
28,64,67			68,69		72	73	74	75,77	28,78
64,67	65		68,69,71		72	73	74	29,30,75,77	29,30,78
35,36, 64,67	65		68,69,71		72	73	74	33,34,35,36, 75,77	33,34,35,36,78
37,38,39,64,67					72	73	74	37,38,39,77	37,38,39,78
40,64,67			40,68, 69,71		72	73	74	40,77	40,78
41,42,43,44,4546, 47,48,64,67			46,48,68, 69,71		72	73	74	41,42,45,46, 48,77	41,42,43,44,45, 47,48,78
49,50,64,67	49,50				72			49,50,75,77	49,50,78
51,52,53,54, 55,64,67			55	55	55,72			55,75,77	51,52,53,54,55,78

and sufficiently copes with the specific problems, and whether the balance of such combination is efficient based on this toolbox. This toolbox can be used to compare management regimes among species or among sites, as well as for conducting inter-temporal comparison analysis on the same species or at the same site (Makino et al. 2011).

9.6 Three Scenarios for the Future of Japanese Fisheries

9.6.1 Senses of Values in Fisheries Policy Discussions

In the analytical processes of policy science, handling senses of values is critically important. Figure 9.1 shows how this point is easily understandable. For example, when certain overfished species need to be allowed to recover, the speed of recovery, or the slope of recovery path, should be decided in the context of the stability of food supply. In this regard, the relative sense of value between (A) and (B) matters. Likewise, who receives what profit (C), to what region the value should be distributed (D), or to what extent noneconomic values such as cultural diversity or beauty of landscapes should be respected (E). This is something that cannot simply be solved by a scientific analysis but must be a matter of public debate.

In addition, what is more important in policy science is that the ingredients of the most desirable combination of measures to realize certain objectives change depending on how the other objectives are treated. For example, if the local/community aspect is given high priority, an efficient approach for (A) would be to strengthen the local authorities in ecological conservation or resource management. If efficiency in monetary terms is prioritized, the market mechanism should be fully utilized, and it becomes essential for the government to apply supplementary policies to prevent market failure, as well as social policies for (D) and (E).

9.6.2 Three Scenarios for the Future of Japanese Fisheries

In the FRA Grand Design Report, three scenarios for the future fisheries in Japan are formulated, referring to the future scenarios in the UN Millennium Ecosystem Assessment (2005). To stimulate public discussion, the three future scenarios are somewhat extreme, since they are designed to clearly highlight the differences in policies arising from different senses of value.

9.6.2.1 Global Competition Scenario: Liberal Scenario Focusing on Economic Efficiency

This scenario clarifies who does what. Specifically, the fisheries policy seeks sustainable economic profit, and other policies cover the rest of the field, as seen in Fig. 9.1. In this scenario, deregulation and free competition are promoted to the maximum extent possible, and the monetary value generated from the industry is maximized. Subsidies to the industry are abolished, and the industry is requested to shoulder their due portion of resources management expenses, such as the cost of resource estimation or monitoring. Although the total number of jobs and vessels will be dramatically decreased, the industry's international competitive power and the employees' income will be increased, and new human resources will also be recruited to fisheries. Management measures such as ITQs that utilize the market mechanism (OECD 2006), and measures centering on economic approaches in Table 9.2 would be compatible with this scenario. However, their introduction is equivalent to the creation of a kind of property rights scheme based on the ecosystem service, and these methods, including their advantages and disadvantages, need to be understood by the public in advance. The collection of royalties (Clark 2006) as an ecosystem service fee is also necessary to prevent excessive investment before the introduction of these measures.

The government, as the resource and ecosystem steward, can rapidly respond to resource and ecosystem fluctuations by changing the TAC or the number of vessels in a top-down way. In principle, the risk of worsening of the economic state of fisheries, which could result from these resource-oriented management measures, would be fully borne by the industry. However, swift control of fishing pressure would make it possible to minimize the risk of resource depletion and maximize the opportunity for resource recovery.

The basic rule of private companies' activity is maximization of profit. They are therefore assumed to be less interested in the public good and nonmonetary values. Also, as a result of pursuing the most efficient production scheme, the variety of fishing operations, equipment, and products would decrease (Matsuda et al. 2008). In addition, if economically rational decisions are to dictate everything, there is the risk of sudden withdrawal from the operation, as well as outflow of the generated values to outside the area, which would destabilize local employment and society. It would therefore be necessary to preempt these problems by applying direct regulatory measures and strict surveillance as social policies by other governmental ministries centering on the administrative approach. Likewise, environmental conservation, local/community maintenance, and stability of food supply should be handled as supplementary policy measures by other ministries in the government.

9.6.2.2 National Food Security Scenario: An Egalitarian Scenario Focusing on the Public Aspects of Food Supply

This scenario focuses on the social role of the fisheries industry as a food production sector. A production system that maximizes fisheries production through resource management and environmental preservation is strategically introduced based on the national government's scientific research. A certain level of income would be guaranteed to all fishers, regardless of the condition of the catch, through strict application of pricing policies (such as collection of fishing taxes or price-supporting policies). In other words, the food supply industry as kind of quasi-public sector would be secured by the national government. People satisfying certain eligibility requirements would be entitled to join the fisheries sector, and the infusion of new personnel would be promoted by adopting a retirement system.

For consumers, public funds would be invested to realize a supply of safe fisheries products at stable prices, insulated from the impact of international supply-and-demand fluctuations. Systematic formation of core fishery areas that match the natural conditions would be conducted to maintain the local economy and local culture. At the same time, data on fisheries operations would be fully disclosed, and monitoring of resources and the environment as well as maintenance of statistical information would be mandatory for fishers.

Since the incentives for innovation or technical progress would be diminished, it is important to maintain and promote operational efficiency by adopting new technologies as they become available. Measures would also be necessary to keep up with consumer needs.

Responding to resources fluctuations is the government's responsibility. For resources that need to be allowed to recover, the catch is determined while considering the stability of the food supply and the speed of resource recovery. The resultant fishery business risk or decision to reduce the number of fishing boats is fully compensated with public funds. Initially, public expenditure would increase, but self-sufficiency would be greatly improved.

In this scenario, measures centering on the administrative and judiciary approaches in Table 9.2 would be introduced into the core measures.

9.6.2.3 Ecological Mosaic Scenario: Regionalist Scenario for Role-Sharing in Conservation of Resources and the Environment

This scenario is based on a role-sharing scheme within the fisheries sector, in which coastal fisheries would play a broader role in regional communities, including taking on a more public role, whereas offshore fisheries would prioritize industrial efficiency.

Coastal fisheries would clarify their accountability for sustainable resource use, ecosystem conservation, and food production depending on the ecological characteristics of each sea area, and play a public role as the core of a regional community. Deregulation would be implemented to allow the entry of new human resources to play public roles, such as interregional distribution of odd lots, establishing a system of local consumption of local products, or autonomous promotion of ecosystem conservation by local people. As a result, a variety of management measures that match the local ecosystem or culture area would be implemented, providing effective protection of nonmonetary values. Maintenance of employment and local life would be highly prioritized, while the efficiency of monetary incentives would be low.

For offshore fisheries, the global competition scenario should be the basis, but it would also be necessary to adopt measures that match the resources and ecosystem of each water area to increase food production as well as enhance the efficiency of their monetary incentives.

Unlike the global competition scenario, responses to resources fluctuation would be conducted through cooperation between the fisheries industry and the government. To ensure a stable supply of marine products to the Japanese population, due consideration would be taken to maintain an appropriate level of fishing operations, and an appropriate resources recovery scenario would be chosen for overfished resources. Closure of fishing grounds and withdrawal from fisheries would then be promoted through reduction in catches and the number of fishing vessels in conjunction with support to compensate for loss of income. In other words, the choice would be to aim at a stable supply of fisheries products and yet

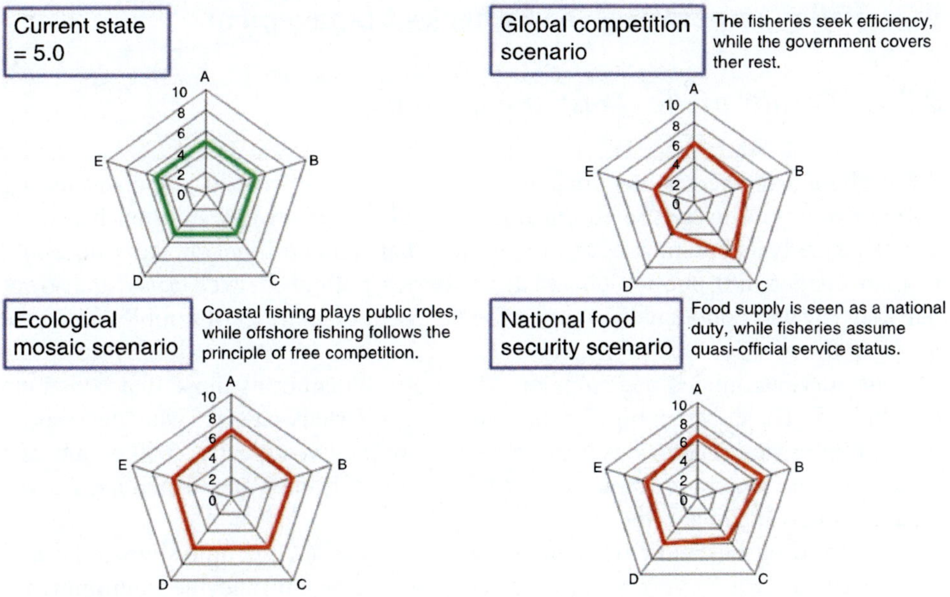

Fig. 9.3 Relative evaluation of three future scenarios (**A**) the resource and environmental policy aspect, (**B**) the food policy aspect, (**C**) the industrial and economic policy aspect, (**D**) the local and community policy aspects, and (**E**) the cultural and scientific policy aspect)

to share the risk of fishery operation deterioration between the government and the industry. Ultimately the risk of no recovery of resources would be higher than that in the global competition scenario.

In this scenario, measures centering on autonomous measures as shown in Table 9.2 would be taken for coastal fisheries, while measures centering on an economic approach would be taken for offshore fishing.

9.6.3 Relative Evaluation of the Three Scenarios

To make a relative evaluation of these policy options, the effects on (A) through (E) of Fig. 9.1 of the three future scenarios are evaluated by the Study Committee members on a scale of 1–10, with the current state set as 5.0. The average evaluation, arranged as a radar chart, is shown in Fig. 9.3. As shown in the figure, each scenario has its particular advantages and disadvantages. To be specific, major improvement, medium improvement, and slight improvement would result from the global competition scenario for (C), (A), and (B), respectively; however, serious deterioration would be expected for (D) and (E). For the ecological mosaic scenario, all of (A) through (E) would expect medium improvement. In the national food security scenario, (B) would see a great improvement, while (A) and (D) would expect medium improvement.

9.7 Policy Imperatives for Fisheries Management

9.7.1 Results of the Questionnaire Survey

As explained in the last section, the sense of value is crucial when discussing policy options. This is something not to be given a uniform solution based on scientific analysis but must be left to public debate and political choice. Therefore, to gain insight into public demand for fisheries policies, a web-based questionnaire survey was conducted in January 2009. The total number of samples analyzed was 2,000.

This survey contains questions on which attributes out of those that constitute (A) through (E), shown in Fig. 9.1, are the most important and asks what the respondent thinks are the relative levels of importance of the five aspects. Another question asks, from the viewpoint of an overall ocean policy framework, which uses of the waters around Japan are important.

For the resources and environmental policy aspect (A), people show a particularly high interest in A-2 (Harmony with the ecosystem and the environment), shown in Fig. 9.1. Particularly high rates come from regions where fisheries production is high, such as Hokkaido, Tohoku (the northern part of the main island), Kyushu, and Okinawa. For the food policy aspect (B), high interest is shown both in B-1 (Increased production and improvement of the self-sufficiency rate) and B-2 (Security of food reliability and safety). This trend is equally seen by region and by gender. For the industrial/economic policy aspect (C), many answers prioritize C-2 (Realization of an efficient and stable operational environment). This trend is higher in younger respondents, namely those in their 20s and 30s, with no major difference seen among regions. For the local/community policy aspect (D), many respondents prioritized D-1 (Job creation for fishing community people). This is more likely to be chosen by the older generation, in their 50s and 60s and older, but falls with decreasing age. By region, it is more likely to be chosen in Shikoku, Tohoku, Kyushu, and Okinawa. Concerning the cultural policy aspect (E), high interest is shown in E-1 (Promotion of fisheries and fishing community culture), particularly in Hokkaido, Kyushu, Okinawa, and Tohoku, where fisheries production is high.

For the ranking of importance among the five policy aspects, (A) through (E), the answer that "all of them are equally important (cannot rank them)" occupied the majority of the answers in all regions, and the national average also went up to 54.5%. Respondents who answered that prioritization is possible (45.4%) were asked about the ranking, and the resultant order of importance is A, B, C, D, and E in descending order.

Finally, the questionnaire asked the respondents to choose what they thought were the most important usages of the waters around Japan (maximum of two choices), and 83.3% of all respondents chose "food production by fisheries," which is followed by "generation of energy from tidal power generation or offshore wind power generation" (54.4%), "transportation" (21.0%), "recreational use" (8.2%), and "creation of space by land reclamation" (1.9%). The respondents who chose "food production by fisheries" were the largest group in every region, and the higher the age bracket, the more such respondents.

9.7.2 Considerations for Future Scenarios

The majority of respondents found it inappropriate to prioritize or weight the five objectives. In the light of this result, of the three future scenarios, we infer that the ecological mosaic scenario, which shows balanced improvement for all principles as in Fig. 9.3, best satisfies public wishes. For the resources and environmental policy aspect (A) considered important among the respondents (45.4%) who said that prioritization is possible, it is also logical to conclude that this scenario, which clarifies coastal fishers' responsibility for using resources sustainably and conserving the environment, is appropriate in the light of the high interest shown by those respondents in A-2 (Harmony with the ecosystem and environment) and that it is more commonly chosen in regions where fishery production is high. Principle 2 of the Ecosystem Approach in the Convention on Biological Diversity states that "Management should be decentralized to the lowest appropriate level" (CBD 2000). This is in-line with the viewpoint found in an increasing number of environmental policies that effective and equitable governance will be best achieved when ecosystem service users in each region implement management in a decentralized and autonomous way (Dolsak and Ostrom 2003).

In the ecological mosaic scenario, the coastal fisheries sector is expected to play a broader role in regional communities, and to take on certain public roles. The real value of a system such as this to society as a whole cannot be accurately evaluated from the viewpoint of fisheries policy only. The most important debate in Japan will be on the role of fisheries in the total framework of the Japanese ocean policy, which includes various sectors and policy aspects. The most desirable ocean policy framework could then be deliberated, in which the fisheries sector, which is supported by 83% of Japanese citizens, is appropriately established and settled. The Basic Act on Ocean Policy of 2007 (Sect. 2.4.3) can act as the legal foundation for an integrated policy framework of this type in the future.

References

Convention on Biological Diversity (2000) Decisions adopted by the conference of parties to the convention on biological diversity at its fifth meeting, UNEP/CBD/COP/5/23

Charles AT (2007) Adaptive co-management for resilient resource systems: some ingredients and the implications of their absence. In: Armtage D, Berkes F, Doubleday N (eds.) Adaptive co-management. University of British Columbia Press, Vancouver, pp 83–104

Clark CW (2006) The worldwide crisis in fisheries. Cambridge University Press, Cambridge

Dolsak N, Ostrom E (2003) The commons in the new millennium. MIT Press, Cambridge

Fisheries Research Agency, Japan (2009) The grand design of resource and fisheries management in Japan. http://www.fra.affrc.go.jp/kseika/GDesign_FRM/GDesign.html. Accessed 30 June 2010

Hilborn R (2007) Defining success in fisheries and conflicts in objectives. Mar Policy 31:153–158

Japan Economic Research Institute (2007) Strategic and drastic reform of fisheries that conserve Japan's fish diet should be expedited. http://www.nikkeicho.or.jp/report/takagifish_teigen_english.pdf. Accessed 30 June 2010

Makino M, Hirota M, Machiguchi Y (2011) An application of the management tool box to coastal fisheries -a case of sea cucumber fisheries. Fisheries biology and oceanography in the Kuroshio 12:25–39 (in Japanese)

Matsuda H, Makino M, Kotani K (2008) Optimal fishing policies that maximize sustainable ecosystem services. In: Tsukamoto T, Kawamura T, Takeuchi T, Beard T Jr, Kaisereds MJ (eds.) Fisheries for global welfare and environment. TERRAPUB, Tokyo, pp 359–369

Millennium Ecosystem Assessment (2005) Ecosystems and human well-being: synthesis. Island Press, Washington, DC

Miyagawa K (1994) Foundation of policy science. Toyo-Keizai-shinbunsya, Tokyo (in Japanese)

Organization for Economic Co-operation and Development (2006) Using market mechanisms to manage fisheries: smoothing the path. OECD Publishing, Paris

Walters CJ, Hilborn R (1976) Adaptive control of fishing systems. J Fish Res Board Can 33:145–149

Willen J (1979) Fisherman behavior and the design of efficient fisheries regulation programmes. J Fish Res Board Can 36:855–858

Chapter 10
Concluding Discussion

Abstract In this book, I have presented six cases of fisheries management (sea cucumber in Mutsu Bay, sandeel in Ise Bay, sandfish in the northern part of the Japan Sea, walleye pollock off the Shiretoko Peninsula, snow crab off Kyoto, and chub mackerel off the Pacific coast). In order to discuss the relevant factors that decide the institutional features of different management regimes, a framework for comparative analysis is presented and applied to six cases. The framework is also applied to the case of ecosystem-based management in the Shiretoko World Natural Heritage Site, and the institutional relationships between the fisheries management cases are discussed. Finally, as challenging viewpoints for future researches, the importance of diversity in institutional paths, differences in management objectives, and resilience to social–ecological changes are discussed.

10.1 Comparison of Fisheries Management Cases

10.1.1 Framework for Comparative Analysis

Fisheries management in Japan is often referred to as a case of co-management of the commons (Berkes et al. 1989; Feeny et al. 1990; McKean 2003; Jentoft et al. 2010). In this book, I have covered six cases of fisheries co-management (sea cucumber in Mutsu Bay, sandeel in Ise Bay, sandfish in the northern part of the Japan Sea, walleye pollock off the Shiretoko Peninsula, snow crab off Kyoto, and chub mackerel off the Pacific coast). To discuss the relevant factors that decide the institutional features of these co-management regimes, a comparative analysis of these six cases is presented in this chapter.

Ostrom (2009) presented a multilevel, nested framework for analyzing outcomes achieved by governance of the commons. In her framework, she set four first-level core subsystems (Resource Systems; Resource units; Governance system; and Users)

M. Makino, *Fisheries Management in Japan: Its institutional features and case studies*, Fish & Fisheries Series 34, DOI 10.1007/978-94-007-1777-0_10,

and 53 examples of second-level variables. This framework is indeed very effective and comprehensive, and applicable to all the fisheries co-management cases presented in this book. In this chapter, however, I would like to propose another framework which is based on the five aspects of fisheries management in Japan (Fig. 9.1), which I hope will better highlight the social and ecological contexts in Japanese cases. The basic structure of my framework is as follows.

As presented in Sect. 9.3, the fisheries sector plays various roles in the Japanese society, and the objectives of fisheries management can be classified into five aspects: (A) resources and the environmental policy aspect, (B) the food policy aspect, (C) the industrial and economic policy aspect, (D) the local and community policy aspects, and (E) the cultural and science policy aspect. Each aspect has related factors or "observable implications" (King et al. 1994) which decide the institutional features of fisheries co-management regime for specific resource and/or fishing type.

As for (A), resources and the environment, the biological and ecological natures of the target species are the relevant factors in management. These include habitat types (coastal/offshore pelagic species, coastal/offshore bottom species, rocky shore species, sandy beach species, etc.), distribution types (local species, widely distributed species, etc.), mobility (sedentary species, migratory species, highly migratory species, etc.), lifespan or maturation age, fecundity, degree of resource fluctuation, prey–predator relationships, functions in ecosystems, trophic level, etc. Relevant information on these factors will be provided by biologists or ecologists. However, the implications for management have been investigated jointly with social scientists.

For (B), the food policy considerations, the characteristics of the resources as food are the key factors. For example, total landing volume, size of demand, economic nature in the food market (relative price, superior goods/normal goods/inferior goods, necessary goods/luxury goods, substitute/complement, price elasticity in terms of landing volume, brand value, etc.), seasonality, locality, usage (for raw fish, for processed fish, for aquaculture feeds, etc.). These factors have been intensively analyzed by fisheries economists and seafood business analysts.

The relevant factors for (C), industrial and economic aspects, concern the fisheries business and their operations. For example, the number of related fishers/vessels, type(s) of fishing gear, fishing grounds, profit ratio, size of capital, degree of economic dependence on a species, degree of business risk aversion, age of fishers, presence or absence of people to take over the business, income level compared to local employee average, stability of income, heterogeneity of these variables within each group, etc. These aspects have been common themes in resource economics (Bjorndal et al. 2007). For example, a famous study by Scot (1993) identified excessive numbers of fishers and their heterogeneity as obstacles to self-governance.

(D) Local and community factors are, for example, the geographical condition of the area (access to large consumption centers, access to good fishing grounds, etc.), rate of job creation in the local economy, complexity of sea surface usage among equipment types as well as among other sectors such as tourism or transport, ripple effects on the local economy via related sectors such as fish processing, fish transportation, shipbuilding, etc. Input–output analysis is a well-known and powerful tool for estimating the overall effects on the local economy (Leung et al. 2001; Heen and Flaaten 2007).

The relevant factors for (E), the culture and science aspects, would be the history of resource use, history of settlement in the community, degree of interpersonal trust among fishers, degree of accumulated information (traditional and scientific) on the target species and fishing grounds, presence or absence of strong leadership and management organizations, etc. The culture or history of resource use has long been a central research topic in cultural anthropology or ethnology (Acheson 1981; Akimichi 1984; Ruddle and Satria 2010). In addition, in recent years, some interesting results have been published in other academic areas. For example, using statistical methods, Ahlerup et al. (2009) have discussed the importance of social capital, in the form of interpersonal trust, for economic development in the absence of formal institutions. Grief (1994) discussed the importance of culture on the nature of organizations, and Prediger et al. (in press) show the impact of culture on cooperation, based on the results of field experiments in common-pool resource management. The social position of the fisheries sector in a community/society can be another important factor in successful management, since it deeply affects leadership, motivation, and the objective-setting processes of the management. For example, Gutierrez et al. (2011) reviewed 130 fisheries co-management cases and identified strong leadership as the most important element in successful co-management.

10.1.2 A Comparison of Six Cases

10.1.2.1 Sea Cucumber in Mutsu Bay (Sect. 4.2)

(A) Resources and the environment: Sea cucumber (*Aposticopus japonicus*) is distributed all along the Japanese coast, from Hokkaido Island to Kyushu Island. It is a typical sedentary species, and Mutsu Bay is a semi-enclosed sea. The local populations in the sea cucumber fishing grounds in Mutsu Bay can thus be regarded essentially as isolated from other populations. This can be an advantageous factor in fisheries management. Although smaller individuals can be sold, it takes about 2 or 3 years to grow to the best market size as well as to mature. This time period is about medium length compared to other fisheries species. There is little information on the prey–predator relationship and resource fluctuations (assumed to be at a medium level).

(B) Characteristics as food: The total numbers of sea cucumber landed are small, but the overall financial yield is very high due to strong recent Chinese market demand for dried sea cucumber and the resultant high prices. This is beneficial in that it generates positive incentives for economically rational management; however, it also attracts poaching. Dried sea cucumber is a typical luxury item. The Kawauchi-machi product has an especially high brand value in the Hong Kong market, so the price elasticity is normally small. This is also an advantage of this species. The fishing season is relatively short (winter only) and raw products are consumed only in winter. On the other hand, the dried products can be stored for years, and exports show no seasonality. To sum up, this species has many advantageous characteristics as food.

(C) Fisheries business and operation: More than 100 small-scale fishers harvest sea cucumber under the auspices of the Kawauchi-machi Fisheries Cooperative Association. Such a large number of fishers would normally result in difficulties in management. However, a good thing is that their homogeneity is very high. Their vessel size, gear used, and income structure are virtually uniform. Their main income in the past was scallop aquaculture, but recently their economic dependency on the sea cucumber resource has been steadily growing, and now almost matches that of scallop aquaculture. There are also sudden high mortality phenomena in scallop aquaculture that occur every few years, making sea cucumber economically important for income stability of fishing people's households. These conditions tend to make fishers more conservative in terms of risk aversion in sea cucumber resource management. The average income of fishers compares well with that of local salaried employees, and the majority of fishers have people to take over the business. This also generates incentives for long-term sustainability.

(D) Local and community: Coastal fishing is one of the most important job sources for local residents. Because most sea cucumber fishers also operate scallop aquaculture, which is another important fishing activity in the bay, and because there are no other large sectors that use the bay, there are no serious disputes over the use of the sea surface. The raw product is consumed locally, and the dried product is exported to Hong Kong, so the distance from the market is not a major problem. These are the advantages of this case. The drying process creates jobs in the local community, but because the total volume (tons) of sea cucumber is smaller than other major processed species, such as sardine, mackerel, or walleye pollock, the ripple effect on the local economy is fairly minor.

(E) History and information: Sea cucumber fishing has a history of hundreds of years in the bay, and local communities have accumulated a great deal of experience and traditional knowledge, on which are based the autonomous measures implemented here. The prefectural research institute scientifically supports resource assessment and resource management activities conducted by local fishers. The fisheries management organization (FMO) organized within the Fisheries Cooperative Association (FCA) has implemented wide-ranging measures under strong leadership. These all tend to contribute to good management.

10.1.2.2 Sandeel in Ise Bay (Sect. 4.3)

(A) Resources and the environment: Sandeel (*Ammodytes personatus*) is widely distributed in Japan, but has fairly low mobility. The local population in Ise Bay inhabits only the inside and around the bay, so the local population is very isolated. The advantageous features of this case are that sandeel take only 3 months to grow to marketable size, and take only 1 year to mature, making it a very fast-growing resource. The results of management measures thus appear quite quickly. However, because this species matures in 1 year, and mortality at the juvenile stage is heavily influenced by natural environmental factors, resource fluctuation is very high. This is a drawback to management. Their prey is zooplankton such as copepods and arrow worms, and their predators are various pelagic and bottom fish, so their trophic level is low.

(B) Characteristics as food: Most human consumption of sandeel is in their juvenile form, and its landing volume (tons) is not very large. The dried juveniles are very popular in Japan, suggesting high market demand and thus high prices. These would be advantageous for management.

Adult-stage sandeel can also be utilized as farm animal and aquaculture feed. Both juveniles and adults are processed before selling. Sandeel processors generally require a reliable supply of adult sandeel, and therefore sandeel fishing people, especially those who target adult sandeel, want a stable catch. This is one of the drawbacks of a fluctuating resource such as this.

(C) Fisheries business and operation: The total number of vessels is very large (about 700), and this makes management difficult. Furthermore, fishers from two different prefectures are catching different sizes of sandeel, which adds to management problems. The fishing gear used by sandeel fisheries is relatively homogeneous in Ise Bay, and this is the only advantage. The economic dependence on this resource is quite high, so the fluctuation in total landings is an issue for fishers' households. Fishers want a steady annual catch, so the stabilization of landings was set as an objective of the Resource Recovery Plan in 2006.

(D) Local and community: The total number of fishers is large, and it occupies an important portion of local job creation. There is a large consumption center (Nagoya) near the fishing grounds. Also, the main aquaculture feed consumption area is the western part of Japan, so the geographical conditions are relatively good for catchers of both juveniles and adults. Both juvenile and adult sandeel is processed before shipping, and it creates ripple effects in the local economy, but it has a short season and the workers at the processing facilities are mainly part-timers. Because the fishing season is short, there is little conflict over sea surface use. There are highly developed heavy industrial sites around Ise Bay and there are many large-scale industrial ships passing through the bay, but the fisheries operations are properly protected.

(E) History and information: There was a long history of conflict between fishers in Aichi and Mie Prefectures, and they experienced a severe resource collapse in the 1980s. A cross-prefectural FMO was newly established in response. Very strong scientific support has been provided to this FMO, and highly scientific management measures have been adopted that have led to successful results. Now both the resource levels and the relationships between fishers in the two Prefectures are satisfactory.

10.1.2.3 Sandfish in the Northern Part of the Japan Sea (Sect. 4.4)

(A) Resources and the environment: Sandfish (*Arctoscopus japonicus*) is a benthic species that migrates long distances, which makes management difficult. However, they gather at specific sites when they spawn in winter. The main spawning area for the northern Japan Sea stock is off Akita Prefecture. The number of eggs (fecundity) is low compared to other species. The average time to maturation is 2 years for females, and their lifespan is about 5 years. These are about average for fish species. Its fluctuation as a resource is large, but the fluctuation mechanism is not yet scientifically clear. They eat amphipods, krill, squid, and fish, but their predators are not well known. In sum, this is a very difficult species to manage.

(B) Characteristics as food: Female sandfish carrying eggs fetch a relatively good price, but they are available only for several weeks in the winter, making them a highly seasonal product. Females in other seasons and males are not attractive as food. The elasticity of price in terms of total catch volume is high, and the price drops markedly if the catch volume is high. It is a traditionally important local food, but the market size is not very large. It is processed in various ways, but in recent years, local demand has been decreasing.

(C) Fisheries business and operation: Both the coastal fisheries (gillnets and set-nets) and offshore fisheries (bottom trawling) in four prefectures are targeting this resource, which means that management is very difficult. Although sandfish fishers catch other species as well, their economic dependency on this resource is fairly high, especially for coastal fishing people. It is worth pointing out that during the autonomous moratorium period lasting 3 years, fishers caught puffer fish and monkfish as alternative target species.

(D) Local and community: As mentioned above, sandfish is consumed mainly by the local market. It is a main target species for a large number of fishers living in this area, so it is an important resource for local job creation, especially for coastal fishers in winter. Sandfish is processed in various traditional ways, but the ripple effect on the local economy appears small. Many households process sandfish at home.

(E) History and information: There is a long history of coastal sandfish fishing, and they are an important local food (actually sandfish is the fish of Akita Prefecture). After the resource collapse in the 1980s, fishers in Akita Prefecture imposed drastic measures for protecting the resource under strong leadership by FCAs and the local government. Although fishers in the other three prefectures share the same stock, there was previously no system for joint management. A new FMO composed of fisheries in four prefectures was formed in 1999, followed by the official management system, the Resource Recovery Plan, in 2003. There is a lot of traditional knowledge of this species, and strong support has been provided by the local research institutes.

10.1.2.4 Walleye Pollock Off the Shiretoko Peninsula (Sect. 8.2)

(A) Resources and the environment: Walleye pollock (Theragra chalcogramma) is a migratory benthic species, but the main spawning ground of the Nemuro Stock is the coastal area off the Shiretoko Peninsula. It takes a relatively long 3–5 years to mature. It also shows large multi-decadal resource fluctuations. These biological conditions are problematic for their management. They are at a relatively high trophic level in the ecosystem. They eat small crustaceans, and are preyed on by marine mammals.

(B) Characteristics as food: The total catch of the Nemuro stock was more than 111,000 tons in 1989, but suddenly dropped to 15,000 tons in 1994. Currently about 9,000 tons are landed. Walleye pollock roe is highly valued in the market as a luxury. The meat is processed as frozen surimi (minced meat), for which there is a large market demand. Some fresh walleye pollock is consumed locally and some is exported to Korea.

(C) Fisheries business and operation: In the 1980s when resource levels were high, there were about 200 gillnet vessels catching walleye pollock around the peninsula. The economic dependency was high, but it has been less high in recent decades with low resource levels. The total income of gillnet fishers has also fallen in recent years. Russian trawlers are also competing for the same resource. The size of vessels and equipment used are fairly standard in Japan, but Russian trawlers tend to be very large vessels, so the overall heterogeneity in the harvest of the Nemuro stock is quite high.

(D) Local and community: Coastal gillnet fishing is one of the most important sources of employment in Shiretoko's local economy. It is distant from major markets, and there are no large fish processing facilities for walleye pollock in this area, so the trickle-down effects on the local economy tend to be minor. The complexity of the sea surface usage is very high, because of the territorial dispute between Japan and Russia, and this causes serious problems in resource management.

(E) History and information: Commercial fisheries developed in this area in the late eighteenth century, so they have a relatively short history. Coastal walleye pollock fishing in this area is even shorter established than other coastal fisheries in other areas. As noted above, walleye pollock shows wide fluctuations, and there was a sudden drop in resource levels in the early 1990s. Various autonomous management measures have been adopted, and the national government set the TAC as the official measure. The walleye pollock FMO is positioned within the local FCA. Strong leadership by core fishers has made the member fishers very eager to implement autonomous management measures based on scientific information, but the resource status is still very low. There is no joint management framework with Russian fishers. Information exchange between fisheries scientists has just begun in recent years.

10.1.2.5 Snow Crab Bottom Trawlers Off Kyoto (Sect. 5.1)

(A) Resources and the environment: Snow crab (*Chionoecetes opilio*) off Kyoto is a sedentary bottom species living in areas deeper than 200 m, and are harvested only by bottom trawlers. Resource fluctuation of the stock appears to be small. These are advantageous characteristics for management. It takes about 5–6 years to mature, so it is a slow-growing species. They eat various species, including crustaceans, fish, squid, and shellfish. Small individuals are eaten by eelpout, flounder, starfish, etc.

(B) Characteristics as food: Because snow crab is a traditional winter food in the Japan Sea area, both demand and prices are high. The total landing in Kyoto Prefecture is not large, but it has brand value. Hard-shell crab is a highly expensive luxury food. On the other hand, the price of soft-shell crab is about one tenth of that of hard-shell crab, and it is commonly available on the market. These conditions generate a strong incentive to manage large hard-shell crab.

(C) Fisheries business and operation: Bottom trawling is offshore fishing, and the capital size of the Kyoto bottom trawlers is larger compared to boats owned by coastal fishers. The total number of snow crab bottom trawlers is small and their vessel size tends to be much the same. More than half of total income is from the snow crab resource, so bottom trawlers' economic dependency is high. These conditions seem advantageous for management.

(D) Local and community: Snow crab is very popular with tourists visiting this area, so the ripple effects from the snow crab to the local economy through tourism sector are large. However, the total number of vessels, as well as the related industries such as processing, is small, and job creation in the fisheries sector is small. There are no other large fisheries or sectors which use the same area as snow crab bottom trawlers, so there are no conflicts as to sea surface use.

(E) History and information: The snow crab FMO in Kyoto has a history of more than 50 years, and bottom trawlers have known each other for generations. They experienced a resource collapse in the late 1970s, and implemented autonomous management in the early 1980s under the strong leadership of the FMO and researchers. The official TAC was also introduced in 1997. Because it is an important species for bottom trawlers and the local economy, high-level scientific information has been gathered by researchers.

10.1.2.6 Chub Mackerel Purse Seiners Off the Pacific Coast (Sect. 5.2)

(A) Resources and the environment: Chub mackerel (*Scomber japonicus*) is an offshore pelagic species. The Pacific stock of chub mackerel is a typical widely fluctuating species in the Japanese EEZ. It shows wide multi-decadal fluctuations and species alternation phenomena with sardine and anchovy.

It migrates extensively along the Pacific coast. These are all drawbacks to effective management. Maturation takes 2–3 years (medium scale). Its prey is zooplankton, anchovy, krill, salp, etc., and its predators are large fish such as sharks and minke whale.

(B) Characteristics as food: The total landing volume of chub mackerel was very large in the late 1970s (more than 1.4 million tons from the Pacific stock), and it is used for human consumption as well as for farm animal and aquaculture feeds. It has always been a popular fish in Japan, and has contributed to increased self-sufficiency. Market demand is large, and the larger specimens are traded at relatively good prices for human consumption. On the other hand, the small individuals fetch only low prices. They are processed for human consumption and for farm animal and aquaculture feed.

(C) Fisheries business and operation: There are many types of fisheries catching chub mackerel off the Pacific coast, but the majority is caught by large- and medium-scale purse seiners. Their economic dependency on chub mackerel is high, although they catch various other species such as sardine, anchovy, and tuna. In the 1980s, the total number of large- and medium-scale purse seiners was about 90 fleets (more than 300 vessels). It is now much smaller at about 25 fleets. On the other hand, the capital size of the large- and medium-scale purse seiners is one of the largest in the Japanese fisheries sector, meaning that it is highly capital intensive. The total gross tonnage of one fleet (three to four vessels) is about 500–600 tons. Their income is high, but costs are also high. For fishing of this type, resource fluctuation in the main target species leads to high fluctuations in income, which generates high risks and high discount rates in fishers' decision making. This is detrimental to management.

(D) Local and community: In the 1970s and 1980s, purse seiners landed large amounts of chub mackerel and sardine at Pacific ports which supported large processing sectors established around the ports. Also, in these periods, new vessels were constructed almost every week, which also supported the shipbuilding sector. The potential job creation and ripple effects are thus very large.

(E) History and information: The history of use of purse seines is shorter than that of other equipment types in Japan. Large- and medium-scale purse seiners are relatively new. The spirit of emulation among fleets is high, and the degree of mutual trust is lower than in other types of fisheries. Because chub mackerel is a popular and important species in the Japanese fish market, it is managed by an official TAC and the Resource Recovery Plan set by the national government; and considerable research effort has been invested. There are several scales of FMOs: at the Prefectural level, the Pacific level, and the national level.

Table 10.1 summarizes the relevant factors in each case, showing clear contrasts among the six case studies. This is only a first step, so most of the discussions are qualitative and include both the causes and results. The next step is to add more case studies, especially of management failure cases, and application of robust analytic methods to this framework.

Table 10.1 Summary comparison of six case studies

	Relevant factors	Sea cucumber	Sandeel	Sandfish	Walleye pollock	Snow crab	Chub mackerel
A: Resource and environment	Mobility	Low	Medium	High	High	Low	High
	Growth time	Medium	Short	Medium	Long	Long	Medium
	Fluctuation	Medium	Large	Large	Large	Small	Large
B: Food character	Price	High	High (juvenile)	High (roe)	High (roe)	High	Medium
	Usages	Medium	Medium	Wide	Wide	Small	Wide
	Demand	Large	Large	Small	Large	Large	Large
C: Business and operation	No. of fishers	Large	Large	Large	Large	Small	Small
	Dependency	High	Medium	Medium	Medium	High	High
	Homogeneity	High	Medium	Low	High	High	Medium
D: Local and community	Job creation	Large	Medium	Medium	Large	Small	Large
	Ripple effects	Small	Large	Small	Small	Large	Large
	Geographic condition	Neutral	Good	Neutral	Bad	Good	Neutral
E: History and information	History	Long	Medium	Long	Short	Medium	Short
	Collapse	No	Yes	Yes	Yes	Yes	Yes
	Leadership	Strong	Strong	Strong	Strong	Strong	Medium

10.2 Discussion on Fisheries Management Cases

10.2.1 The Roles of Fishers, Science, and Governments

There are many types and forms of co-management regimes, such as from "instructive" and "consultative" to "informative" (Berkes 1994), or "government based," "market based," or "community based" (Copes and Charles 2004). Each co-management regime has specific features as to role sharing among relevant actors.

Local fishers have a strong incentive to autonomously manage sedentary species and locally independent populations that fetch high prices, and they play a central role in their management. Sea cucumber in Mutsu Bay and snow crab off Kyoto are typical cases. On the other hand, for highly migratory species or widely distributed species caught using various types of equipment and fishers from different communities, management measures imposed by the government are central. A typical example is that of chub mackerel off the Pacific coast. Hiroyoshi and Sano (1998) pointed out, based on the review results of 58 case studies of fisheries management in Japan, that the role of government and the federation of FCAs are the key to managing widely distributed resources. Note that, even for official government measures such as TACs or Resource Recovery Plans, fishers' organizations play an important role in the planning and implementation processes. As explained in Chap. 2, "Fisheries management by resource users themselves" has been the fundamental concept that governs fisheries management. In that sense, the abilities of fishers are critically important to good management. Providing infrastructures for increasing ability, such as education, extension officers, fishing mentor systems, etc. (Sects. 1.4 and 1.5), are important government roles.

One interesting point is the gradual development of management regimes in the case of sandfish. Management had been launched autonomously by the local fishing people, but, in a step-by-step manner, they implemented a wider-scale and more official management framework, and established a high-quality finished form of management regime with successful results. As this case shows, another important government role is the coordination of heterogeneity in equipment type and capital size. Of course, for resources which closely relate to international disputes such as the case of walleye pollock in the Shiretoko area, the role of the government is crucial.

Interesting differences can also be found among the management regimes for three migratory species (sandfish, walleye pollock, and chub mackerel). Sandfish migrate widely over four prefectures but gather at a critical period (spawning) in small areas off Akita Prefecture. In this case, the fishers in the spawning area had already taken leadership and begun rigid management, and that type of management became the foundation for a later management regime that covered all the relevant prefectures and equipment types. Similarly, walleye pollock gather for spawning in areas off the Shiretoko Peninsula, and the local gillnet fishers set those areas as MPAs. However, due to the territorial dispute, the management regime cannot be extended to cover Russian fishers and related fishing grounds. The fishers are thus not able to play a central role in management: the government must take the lead. Note that the price of roe is high both for sandfish and walleye pollock. On the other hand, the Pacific stock of chub mackerel spawns in coastal areas off the central and western coasts of Japan, but the major fishing grounds are the northern parts, so there is a spatial discrepancy. Also, unlike the sandfish or walleye pollock, the price of chub mackerel in the spawning period (winter to spring) is very low. As a result, the management measures lean more toward a government-based regime than those for sandfish and walleye pollock.

There are several other points to which the government needs to pay attention. For resources that can provide a small incentive for fishers to self-manage but generate large ripple effects on the local economy, or have special importance in the national food security or local culture, the government should intervene in management to an appropriate degree. For example, chub mackerel has major ripple effects on landing areas, but high uncertainties generated due to high resource fluctuations and migration impede self-management. This situation creates one more reason to advise government-oriented management measures for this resource. The government should also play a central role for resources on which local fishers living in remote areas are economically dependent, but which have numerous negative factors for self-management.

Science is playing an essential role in all of these cases. Scientific information can bring a common awareness of the need for the management of species with high uncertainties or high fluctuations (sandeel in Ise Bay, sandfish in the northern Japan Sea, and chub mackerel off the Pacific coast). Snow crab is a sedentary species with very high price, so there is a strong incentive for autonomous management, but this species takes time to grow to marketable size and to mature. At the beginning, as detailed in Sect. 5.1, fishers were very skeptical of the effectiveness of MPAs. Therefore, in this case, scientific information provided by the prefectural research institute was critical to the adoption of MPAs and the beneficial results.

Also, as discussed in Sect. 2.4, the objective evidence for sustainable resource use is generally weak for autonomous management measures. Science can help counteract this weakness. Providing scientific advice for regular validations and adaptive revisions for autonomous management measures is another important role of science.

Modern scientific knowledge generally requires a great deal of social capital (research facilities, statistical systems, human resources, etc.,) and has high running costs, so it cannot be applied to all relevant species. Fishers' knowledge, experience, and information are very valuable and should be fully utilized. However, they are basically biased by economic incentives, and their view is limited to the scale of their own fishing grounds. Another role of science is thus to counteract these biases and integrate them to scale up to larger time and spatial scales. Finally, as noted in the Shiretoko case (Chap. 8), science can provide suggestions for solving problems in fisheries and ecosystem-based management over territorially disputed areas.

Table 10.2 Major target species of Fisheries Management Organizations (FMOs) in 2008 (Source: MAFF 2010)

Species	Number of FMOs	Ratio to the total number of FMOs (1,738) (%)
Abalone (*Haliotis* spp.)	594	34.2
Spiny top shell (*Batillus cornutus*)	439	25.3
Sea urchin (*Echinoidea*)	428	24.6
Sea cucumber (*Holothroidea*)	324	18.6
Bastard halibut (*Paralichthys olivaceus*)	318	18.3
Silver sea bream (*Pagrus major*)	214	12.3
Righteye flounder (*Pleuronectidae*)	207	11.9

Note: Multiple species are managed simultaneously by single FMOs

10.2.2 Fisheries Cooperative Associations and Fisheries Management Organizations

Under the fisheries management regime in Japan, Fisheries Cooperative Associations (FCAs) and Fisheries Management Organizations (FMOs) play the most important roles, so they embody a type of social capital (Pretty 2003). As explained in Sect. 2.3, FCAs are composed of local fishing people, and are rooted in each fishing community. There were 1,041 FCAs in 2008 (MAFF 2010). On the other hand, an FMO constitutes a group of fishers who use the same equipment or target the same species. They are usually organized within an FCA, but sometimes FMOs include members from several neighboring FCAs or even from FCAs in several prefectures. FMO rules are more detailed and stricter than FCA regulations (Table 2.2).

There have been many research studies on FCAs and FMOs by Japanese researchers. For example, Yamamoto (1989) pointed out that FCAs and FMOs play a central role, both in planning and controlling fisheries operations aiming at efficient and sustainable resource use, and in distributing the wealth from fishing grounds in an equitable and fair way among the members of the organizations. Lou (1996) emphasized the leadership of the heads of FCAs or FMOs, such as fostering trust between fishers, or coordination capacity, etc., as an essential factor in successful management.

Several factors listed in Table 10.1 can be discussed using the statistics on FMOs in the Fisheries Census of 2008 (MAFF 2010). There are 1,738 FMOs in Japan, a 13.4% increase since the last census in 2003. Most FMOs are managing high-price sedentary resources such as abalone, sea urchin, sea cucumber, etc. (Table 10.2). Finfish such as bastard halibut, sea bream, and righteye flounder are the next most popular subject species. These fish also fetch high prices and have fairly low mobility.

The distribution of FMO size is shown in Table 10.3. There is a wide variety seen, with no clear tendency. However, according to a statistical breakdown of FMO sizes in terms of species caught and equipment used, FMOs which manage sedentary

Table 10.3 Size categories in Fisheries Management Organizations (FMOs) in 2008 (Source: MAFF 2010)

Number of fishers	Number of FMOs	Ratio to the total number of FMOs (1,738) (%)
Fewer than 10	244	14.0
10–20	356	20.5
20–30	224	12.9
30–50	315	18.1
50–100	290	16.7
100–200	167	9.6
200–300	62	3.5
More than 300	80	4.6

resources such as abalone, spiny top shell, sea urchin, and sea cucumber tend to have a medium size (30–100 fishers), while finfish FMOs are composed of larger numbers of fishers. The reason is probably that because the FMO should ideally cover all related fishers harvesting the same resource, and finfish have higher mobility than sedentary species, fishers targeting finfish are spread over a wider area and therefore are larger in number.

As for type of fishing gear, bottom trawlers and gillnet fishers tend to form small FMOs (fewer than 50 fishers), while FMOs for "pole and line fishing" and "harvesting shellfish and seaweed" tend to have larger numbers. This appears to be because the necessary capital size for the latter type is smaller and the number of related fishers is larger (and presumably older on average).

10.2.3 Issues of Scale

Space and timescale are crucial factors for successful fisheries management. For spatially small and sedentary resources which can be managed by a single FMO, such as sea cucumber in Mutsu Bay and snow crab off Kyoto, the FMO members can easily benefit from the effects of their management efforts. On the other hand, for highly mobile and migrating species such as sandfish in the northern Japan Sea and chub mackerel off the Pacific coast, there are major externalities. For such resources, appropriate systems need to be established to cover the migrating area and all the relevant fishers harvesting the resource. The lack of an appropriate spatial scale in fisheries management could lead to management failure (Cardinale et al. 2010).

In Japanese fisheries management, there are several levels and scales of official organizations (Table 2.2), extending from the community level (FCAs) or prefectural level (Area Fisheries Coordinating Committees) to the ecosystem level (Wide-area Fisheries Coordinating Committees) and national level (Fisheries Policy Council). No less important are FMOs with various scales in space. For example, large- and medium-scale purse seiners have organized federations at prefectural level, Pacific level, and national level. Sandeel fishers in Ise Bay have also organized a multi-prefecture FMO called the General Assembly of Fisheries Unions for Sandeel Fishers in Ise Bay (Fig. 4.6). Fishers targeting the northern Japan Sea stock of sandfish belong to local FMOs which are organized within each FCA, a prefectural federation of such FMOs, and a stock level federation which covers all fishing gear types in four prefectures, called the Sandfish Resource Measures Council (Fig. 4.9).

According to the Fisheries Census of 2008 (MAFF 2010), there are 141 FMOs covering multiple municipalities (8.1% of the total) but only 11 (0.6%) covering multiple prefectures. As these statistics show, cross-jurisdictional or prefectural management is still rare in Japan, but the development process of sandfish management regime shows a good way forward for the management of highly migratory and widely distributed species. As explained above, they have implemented various scales of management measures with different official levels in a step-by-step manner.

Next is timescale. As far as the Japanese fisheries management is concerned, I think there are at least three levels of timescale that are of critical importance. The first and the longest is the 20–30 year scale. This roughly corresponds to the period between generations of fishers and the lifespan of fishing vessels. The second is a 5 year period. From the legal viewpoint, it corresponds to the duration of fishing rights and licenses. From the viewpoint of ecological systems, even long-lived species such as crab, tuna, large benthic species, etc., can mostly mature within 5 years. At least one strong year class can also be expected for widely fluctuating species such as chub mackerel, sardine, or walleye pollock. Finally, the shortest but not the least important timescale is, of course, 1 year. This timescale is especially important for business operations. As mentioned in Sect. 5.2, bankruptcy is a serious event in Japanese culture (it is perceived as almost equivalent to the death penalty), and therefore the yearly settlement of business accounts is very important. Also, for short-lived species, yearly fluctuations tend to be wide, such as the sandfish case presented in Sect. 4.3. In devising quantitative models for fisheries management, these three levels of timescale (1, 5, and 20–30 years) need to be clearly incorporated, and the consistency of management measures in terms of these three levels of time period should be analyzed.

Another important issue relating to timescale is time lags. Depending on the biological and ecological nature of the target species, there is often a considerable time lag between the adoption of management measures and the emergence of management effects, and this time lag sometimes generates a high risk and therefore a high discount rate in fishers' decision making, which stands in the way of long-term thinking (Makino 2007). Of the case studies in this book, walleye pollock and snow crab fit these conditions. For these two species, science should play the vital role to reduce the uncertainties and risks in long-term fisheries management. A great deal of scientific effort has, in fact, been devoted to these species by researchers at both national and local fisheries research institutes as well as by university-based researchers. When the time lag between management implementation and payback is too long and/or the associated risks are too high, fishers and the local economy are unable to make biologically correct decisions. For species that need fast and secure resource recovery, the government should play a role, especially in compensating for economic risks (see Sect. 2.4, Economic Support System for the Resource Recovery Plan).

10.3 Expansion of a Comparative Framework for Ecosystem-Based Management

In this section, I would like to try to expand the above framework for fisheries management to the ecosystem-based management, and apply it to the Shiretoko World Natural Heritage case I presented in Chap. 8.

To be able to apply it to the Shiretoko case, I have had to make some small modifications to the original framework. First, (A) resource and environment should be (A'), ecosystems, which summarizes the natural scientific features of the ecosystems. (B') would concern human appreciation of "ecosystem services," which include provisioning services (food, freshwater, wood and fiber, fuel, etc.); regulating services (climate regulation, flood regulation, disease regulation, water purification, etc.); cultural services (aesthetic, spiritual, educational, recreational, etc.); and support services (nutrient cycling, soil formation, primary production, etc.) (Millennium Ecosystem Assessment 2005). Likewise, (C') would describe the economic activities and sectors using the above ecosystem services. (D') is regional factors relating to ecosystem services, similar to the original (D). For example, geographical conditions, rate of job creation, complexity of ecosystem service usages, ripple effects, etc. Finally, (E') is factors relating to the history of ecosystem service uses, social capital, traditional/scientific information about the ecosystems, the presence of management organizations, etc.

As for (A'), ecosystems, the Shiretoko ecosystem is a subarctic ecosystem. It is the southernmost limit of seasonal ice floes in the northern hemisphere. There is a close relationship between coastal and terrestrial ecosystems, but at the same time, the Shiretoko ecosystems are assumed to form a part of the huge marine ecosystem that includes the northern territories and Russian territories.

Of the many ecosystem services (B') in Shiretoko, the provisioning services (fisheries resources) and cultural services have been the most appreciated and utilized by the social system. The ice floe also supports the high productivity of the area (supporting services). These are appreciated as having "outstanding universal value" and therefore this area was added to the UNESCO World Natural Heritage List in 2005.

Based on the above ecosystem services, two economic sectors (C') have been developed: the fisheries sector and the tourism sector. The dependency of these sectors on ecosystem services is quite high, making the conservation of the ecosystem structures and functions crucial to these two sectors. The above two sectors (fisheries and tourism) are the main sectors in the local economy and community (D'), and the main sources of employment for local people. The trickle-down effects from these two sectors to the local economy are also significant. There are sometimes conflicts about the usage of ecosystem services between fisheries and tourism, as explained in Sect. 8.3.5.

History and information (E') factors are summarized as follows. This area has been inhibited by people for thousands of years, and it has been a famous fisheries production area. The peninsula was designated as a National Park in the 1970s, and attracts a lot of tourists. After its addition to the Heritage List, intersectoral coordination for ecosystem-based management has rapidly progressed under a newly established system (Fig. 8.2). Since the end of the Second World War, there have been territorial disputes in the adjacent area. Although there is clearly a close relationship between the ecosystems in Shiretoko and the disputed areas, there are no frameworks for large-scale ecosystem conservation.

In this book, I give only one case of ecosystem-based management, but when other ecosystem-based management cases are summarized based on the above framework and compared to each other, many interesting discussions can take place. Notably, a comparison with the southern part of Japan (a tropical marine ecosystem) promises useful insights. On the other hand, when compared to the fisheries management cases discussed in the last section, the ecosystem-based management in this Shiretoko case has several management features.

First, it is clear that there is more variety in the types of ecosystem service uses than there are fisheries management cases; and in response to that variety, there is a variety of management regimes. The point here is that the variety of management regimes is not like the cascade structure found in the case of the sandeel in Ise Bay or sandfish in the northern Japan Sea, but rather a mosaic system with diverse types of management regimes for various types of uses. In the Shiretoko case, there is no umbrella organization which covers all uses. The newly established coordinating system (Fig. 8.2) is a system, or network, of related actors and management authorities. Therefore, role sharing among ecosystem service users, government, and science also takes the form of a mosaic. For example, tourism sector management is relatively top down, since tourists (cultural service users) are highly heterogeneous and widely dispersed. On the other hand, fishers (provisioning service users) play a core role in fisheries management as well as in ecosystem monitoring. Science is important for understanding ecosystems and reducing the uncertainties related to ecosystem services, in the same way as in fisheries management cases. Scientists also play a core role in drafting management plans and balancing various interests among sectors. In addition, it is expected that scientists will play a leading role in sharing ecosystem information between the heritage area and the territorial dispute area. Therefore, in the Shiretoko case, co-management regime appears to resemble a "science-based co-management mosaic." It might be because this was the first case of cross-sector ecosystem-based management in Japan, and thus the theoretical aspects took the lead in designing the management regime.

In the scale issue discussions on fisheries management, I have emphasized the importance of the appropriate scale of FMOs. In the case of the ecosystem-based management of the Shiretoko World Natural Heritage site, the analogy could be applied to the sector issues, i.e., to emphasize the importance of an appropriate range of sectors being included in the coordinating system. Without such a system, cross-sector coordination never emerges. Of course, leadership by government officers and leading scientists were also crucial to making this new system work well, as discussed in Sect. 8.5.

Finally, I would like to discuss the triggers for changes in the management regimes. In most of the fisheries management cases presented in this book, severe resource collapses prompted the implementation of rigid management measures. In other words, the difficult experiences during the collapse changed the outlook of the fishing people affected, who then went on to change their management regime. Similarly, in the Shiretoko case, nomination to the World Natural Heritage List in 2004 was the trigger for changes in the thinking of the people involved. According to the officer of the Minister of the Environment who was in charge of the Heritage matter, he was convinced that a simple combination of conventional environmental conservation and fisheries management would not be enough to persuade UNESCO and IUCN. He actually felt compelled to do something innovative to achieve inclusion in the Heritage List. This was, to my understanding, an important trigger that led to the new regime in Shiretoko.

Therefore, just as external political factors influenced the process of the Fisheries Reforms after the Second World War (Sect. 2.3) and ultimately led to the adoption of TACs in 1996 (Sect. 2.4), external factors such as international organizations or NGOs influenced the management regime in Shiretoko. This is where I am struck by the importance of the "social, economic and political settings (Ostrom 2009)," which I could not incorporate clearly into my framework. At the same time, a no less important point is that, just as the Japanese-style TAC enforcement system was created in order to ratify the UNCLOS, the Japanese-style ecosystem-based management system was created in a bid to achieve inclusion in the World Natural Heritage List. They are not a copy of other regimes in other countries, but original Japanese regimes based on existing institutional frameworks. There is no "one-size-fits-all" regime, but there are clearly diverse institutional paths toward achieving the objectives, as discussed in the next section.

10.4 Future Challenges

There are many other important topics for fisheries management which I did not cover in this book, such as the role of the processing and transporting sectors, aquaculture, and freshwater fisheries. Also, there is no doubt that consumers are taking on an expanding role in fisheries and ecosystem-based management in Japan. Three more research topics are discussed as the final section of this book.

10.4.1 Diversity in Institutional Features and Their Historical Paths

Just as in ecological systems, there is diversity in social systems and institutions. Research on ecological diversity seems far in advance of research on social–institutional diversity, but I think that common analytical viewpoints and similar quantitative methods can to some extent be applied to both ecological and social systems studies.

As discussed at the end of Chap. 8, different ecological backgrounds and different institutional backgrounds naturally follow different paths to ecosystem-based management. A general theory which explains such diversities in trajectories will be only developed by engaging in comparative case studies. More case studies are needed on ecosystem-based management. Fortunately, Japan has a wide variety of marine ecosystems, from the subarctic to the tropical. Case studies in different ecosystems within Japan and a comparison of cases will be the next task.

Rights-based management (Neher et al. 1989; Millennium Ecosystem Assessment 2005; Libecap 2010) is attracting attention in discussions on fisheries management. I completely agree with the importance of rights-based management. An examination of this topic from the viewpoint of ecosystem-based management shows that the specific nature of rights is gaining in significance. For example, the difference between area-based fishing rights and species-based fishing rights will create essential differences in the resultant ecosystem-based management regimes. Generally speaking, area-based measures are likely to be more compatible with ecosystem-based management, especially in coastal areas.

Similarly, differences in management philosophy or the principal management tools will inevitably constrain the feasible institutional paths. There are many social–ecological factors that will explain existing differences in management philosophy or tools among countries and areas. For example, tropical ecosystems with high biodiversity might result in input control-oriented management measures as the best solution, while arctic ecosystems may be managed by output control measures. In this sense, as emphasized in Chaps. 6 and 8, I believe that a deep understanding of existing institutional backgrounds, as well as comparisons with different institutions and ecosystems, is a prerequisite for a better and smoother transition to an optimal ecosystem-based management in any country or area.

10.4.2 Objective(s) of Management

The objective(s) of management are closely related to the social backdrop. In Japan, for example, people have been living in these islands for thousands of years, and have eaten seafood as their principal source of animal protein for an equally long time. Japanese citizens regard the fisheries sector to be important as a provider of seafood (see the results of the questionnaire survey in Sect. 9.7). As exemplified by the case of the Shiretoko World Natural Heritage Site (Chap. 8) and the Sato-umi initiatives (Sect. 7.5), local fishing is an integral component of local ecosystems, rather than a threatening intrusion into "pristine ecosystems." Therefore, as shown in the cases of autonomous marine protected areas (AMPAs) in Sect. 7.3, local fishers play a core role in local ecosystem conservation activities, and public citizens are positively participating in such activities.

The leader of the environmental NGO participating in the seagrass bed MPA activities in Yokohama Bay said, "Even with very wide gaps in values or beliefs among participants, I think we can at least share the importance of education for local children… in that sense, it is important to give children the experience of catching and eating seafood in the field." (Citizen's Group Meeting for the Promotion of Nature Restoration 2005). I think these words clearly highlight the relationship between marine social and ecological systems in Japan, and are likely to be a good basis for the objective setting of goals both in fisheries management and in ecosystem-based management in Japan. Other countries or areas will have other bases and other objectives.

In Chap. 9, I summarized five aspects of the objectives of the Japanese fisheries policy, but I did not provide enough analysis on why these five aspects have developed in the Japanese society. This topic needs to be developed more and international comparisons should be made with fisheries policies in other countries, for example, with the Anglo-American Law countries, South-East Asian countries, and African coastal countries. I believe such comparisons will bring fruitful insights to international discussions on the objectives of fisheries management as well as ecosystem-based management.

10.4.3 Changes in Social–Ecological Systems and Resilience

The 15th Conference of Parties of the United Nation Framework Convention on Climate Change was held in December 2009 at Copenhagen. At this conference, the first "Ocean Day" was convened and the impacts of climate change on the oceans were intensively discussed. The expected effects include changes in sea surface temperature (SST), sea level, frequency and magnitude of weather events such as storms and typhoons, decadal climate oscillations, and ocean acidification levels. Some of the expected effects from such changes include deterioration of marine ecosystems, and changes in ocean productivity and species distributions (Brander 2010).

The western Pacific area, in which Japan is located, is a world biodiversity hotspot. Sharp rises in the sea surface temperature (SST) have been observed in this area. For example, in the Kuroshio Current area, the speed of the SST rise in the last 100 years is about 1.5–3 times higher than the world average. The amount of ice floes around the Shiretoko Peninsula is reportedly decreasing (Sect. 8.5), which may trigger significant changes in the coastal marine ecosystem. As the International Council for Science (ICSU) and the International Social Science Council (ISSC) have pointed out, the "Grand Challenges" are better integration and cooperation between the social sciences and biophysical sciences (Reid et al. 2010). Our area of direct influence is human society, not ecosystem structures or functions. All the issues in fisheries and ecosystem-based management are thus essentially people-oriented issues. There is a great need for input from the social sciences.

Most fisheries management case studies show that after local fishers experience resource collapses, they change their approach and implement strict management measures. The key here is that such changes happened before the situation crossed an ecological threshold and became irreversible. In other words, they changed or improved their management while the resource level was still recoverable. Natural sciences was a major locomotive for this change, since fishers are not always aware of the full ecological picture. Also, in the Kawauchi-machi FCA case at Mutsu Bay, where sea cucumber fisheries and scallop aquaculture are the two main sources of income, there were very serious mass mortality accidents of cultured scallops in 2003 and 2010. As an emergency measure to stave off bankruptcy, the fishers increased their sea cucumber catch in these years. As a result, the Kawauchi-machi FCA members suffered much less economic damage than other fishers in Mutsu Bay. This is an empirical case showing that the adoption of good management for one resource can increase overall economic resilience to sudden collapses in other resource levels. Likewise, to mitigate the effects of climate change, an appropriate combination of resources and harvesting strategies could be quantitatively analyzed from the viewpoint of the vulnerability of fishing communities.

References

Acheson JM (1981) Anthropology of fishing. Annu Rev Anthropol 10:275–316

Ahlerup P, Olsson O, Yanagizawa D (2009) Social capital vs institutions in the growth process. Eur J Polit Econ 25:1–14

Akimichi T (1984) Territorial regulation in the small-scale fisheries of Itoman, Okinawa. In: Ruddle K, Akimichi T (eds.) Maritime institutions in the Western Pacific (senri ethnological studies 7). National Museum of Ethnology, Suita

Berkes F (1994) Property rights and coastal fisheries. In: Pomeroy RS (ed.) Community management and common property of coastal fisheries in Asia and the Pacific: concepts, methods and experiences. International Center for Living Aquatic Resources Management, Metro Manila, pp 51–62

Berkes F, Feeny D, McCay BJ, Acheson JM (1989) The benefits of the commons. Nature 340:91–93

Bjorndal T, Gordon DV, Arnason R, Sumaila UR (2007) Advances in fisheries economics: festschrift in honour of professor Gordon R. Munro. Wiley-Blackwell, Oxford

Brander K (2010) Climate change and fisheries management. In: Grafton RQ, Hilborn R, Squires D, Tait M, Williams M (eds.) Handbook of marine fisheries conservation and management. Oxford University Press, Oxford, pp 123–136

Cardinale M, Bartolino V, Llope M, Maiorano L, Skold M, Hagberg J (in press) Historical spatial baselines in conservation and management of marine resources. Fish Fish.

Citizen Group Meeting for the Promotion of Nature Restoration (2005) Nature restoration activities linking forests, sato (villages), rivers, and ocean: lessons from 13 advanced cases in Japan. Chuohoki Publishers, Tokyo (in Japanese)

Copes P, Charles A (2004) Socioeconomics of individual transferable quotas and community-based fishing management. Agric Resour Econ Rev 33:171–181

Feeny D, Berkes F, McCay BJ, Acheson JM (1990) The tragedy of the commons: twenty-two years later. Hum Ecol 18(1):1–19

Grief A (1994) Cultural beliefs and the organization of society: a historical and theoretical reflection on collectivist and individualist societies. J Polit Econ 102:912–950

Gutierrez NL, Hilborn L, Defeo O (2011) Leadership, social capital and incentives promote successful fisheries. Nature, Jan 5. [Epub ahead of print]

Heen K, Flaaten O (2007) Spatial employment impacts of fisheries management – a study of the arents Sea and the Norwegian Sea Fisheries. Fish Res 85:74–83

Hiroyoshi K, Sano M (1998) Features and issues in the resource management-type fisherie. Zengyoren, Tokyo (in Japanese)

Jentoft S, McCay BJ, Wilson DC (2010) Fisheries Co-management: Improving Fisheries Governance through Stakeholder Participation. In: Grafton RQ, Hilborn R, Squires D, Tait M, Williams M (eds.) Handbook of marine fisheries conservation and management. Oxford University Press, Oxford, pp 675–686

King C, Keohane RO, Verba S (1994) Designing social inquiry: scientific inference in qualitative research. Princeton University Press, Princeton

Leung P, Heen K, Bardarson H (2001) Regional economic impacts of fish resources utilization from the Barents Sea: trade-offs between economic rent, employment and income. Eur J Oper Res 133:432–446

Libecap GD (2010) Allocation Issues in Rights-Based Management of Fisheries: Lessons from Other Resources. In: Grafton RQ, Hilborn R, Squires D, Tait M, Williams M (eds.) Handbook of marine fisheries conservation and management. Oxford University Press, Oxford, pp 572–581

Lou X (1996) The organizational nature and abilities in fisheries management organizations. Reg Fish Res 37(1):51–71 (in Japanese)

Makino M (2007) A risk analysis of adaptive fisheries management. Jpn J Fish Econ 52(2):49–67 (in Japanese)

McKean M (2003) Common-pool resources in the context of Japanese history. World Wide Bus Rev 5(1):132–159

Millennium Ecosystem Assessment (2005) Ecosystems and human well-being: synthesis. Island Press, Washington, DC

Ministry of Agriculture, Forestry and Fisheries (2010) Fisheries census of 2008. Ministry of Agriculture, Forestry and Fisheries, Tokyo (in Japanese)

Neher PA, Arnason R, Mollet N (1989) Rights based fishing. Kluwer Academic Publishers, Boston

Ostrom E (2009) A general framework for analyzing sustainability of social-ecological systems. Science 325:419–422

Prediger S, Vollan B, Frolich M (2011) The impact of culture and ecology on cooperation in a common-pool resource experiment. Ecol Econ 70:1599–1608

Pretty J (2003) Social capital and the collective management of resources. Science 302:1907–1912

Reid WV, Chen D, Goldfarb L, Hackmann H, Lee YT, Mokhele K, Ostrom E, Raivio K, Rockstrom J, Schellnhuber HJ, Whyte A (2010) Earth system science for global sustainability: grand challenges. Science 330:916–917

Ruddle K, Satria A (2010) Managing coastal and Inland Waters. Springer, London

Scot A (1993) Obstacles to fishery self-government. Mar Resour Econ 8:187–199

Yamamoto T (1989) Role of fisheries cooperative associations in the resource management-type fisheries. Jpn J Fish Econ 33(2–3):40–61 (in Japanese)

Index

M. Makino, *Fisheries Management in Japan: Its institutional features and case studies*, Fish & Fisheries Series 34, DOI 10.1007/978-94-007-1777-0,